T0145930

PUBLICATIONS OF THE ISRAEL ACADEMY

OF SCIENCES AND HUMANITIES

SECTION OF SCIENCES

———

FAUNA PALAESTINA

INSECTA VI — TRICHOPTERA OF THE LEVANT

– IMAGINES –

FAUNA PALAESTINA

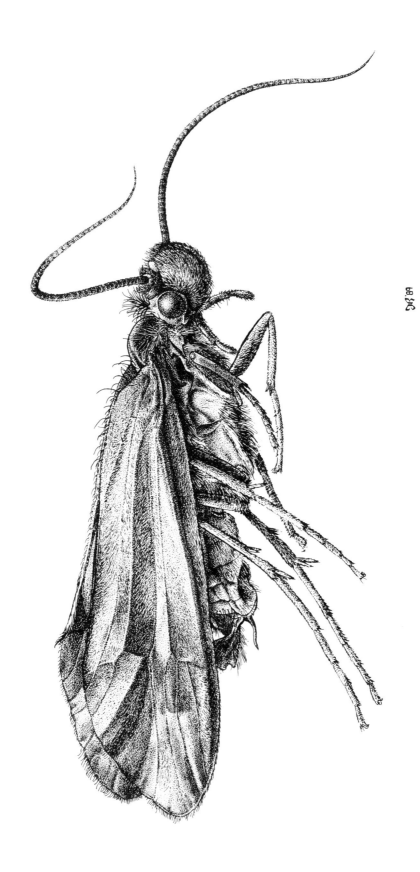

Sericostoma flavicorne Schneider, 1845, male

FAUNA PALAESTINA · INSECTA VI

TRICHOPTERA OF THE LEVANT

– IMAGINES –

by

LAZARE BOTOSANEANU

Jerusalem 1992

The Israel Academy of Sciences and Humanities

Author's Address:
Institute of Taxonomic Zoology
University of Amsterdam
Plantage Middenlaan 64
1018 DH Amsterdam, The Netherlands

ISBN 965-208-013-6
ISBN 965-208-098-5

Printed in Israel
at Keterpress Enterprises, Jerusalem

CONTENTS

PREFACE, ACKNOWLEDGEMENTS

The study of the Israeli and Levantine caddisflies has occupied my thoughts since 1963, when I published the first results. I am indebted to Prof. F.D. Por (Jerusalem) who invited me many years ago to prepare this Fauna volume.

My thanks are expressed to various colleagues in the Departments of Zoology of The Hebrew University of Jerusalem (especially Prof. F.D. Por and members of staff of the Inland Water Ecological Service [I.E.S.], Dr R. Ortal in particular) and of Tel Aviv University (especially Prof. J. Kugler) for help offered during my two collecting trips to Israel and Sinai (1971 and 1984).

With quite a few exceptions, the illustrations appearing in this book are drawings made by myself. The habitus drawings are the work of Mr J. de Rond (Lelystad, The Netherlands). Some of the illustrations, already printed in several publications of mine, were inked by Mrs Violeta Berlescu-Salcher (Bucharest) or by Mr J. Zaagman (Amsterdam). However, the overwhelming majority of drawings were inked by Mr D.A. Langerak (Amsterdam). Miss Annelies Stoel (Amsterdam) typed the manuscript. Work has been done on the English styling by Mr R. Amoils (Jerusalem). Ms Ilana Ferber and Dr Heather Bromley (Jerusalem) did the scientific editing; Ms Ferber also compiled the index and saw the book through the press. To all of them, my best thanks are extended.

My sincere thanks to Dr Ch. Dimentman (Jerusalem) who competently and kindly answered many questions concerning names and other data pertaining to Israeli and other localities.

A number of persons helped with various specific problems. I hope that their assistance has always been correctly acknowledged in the text of the present book.

Most of the specimens studied for this Fauna volume are deposited either in the National Entomological Collection at Tel Aviv University's Zoological Museum, in the Zoological Museum of the University of Amsterdam, or in the National Aquatic Invertebrate Collection at The Hebrew University of Jerusalem. Maybe other specimens are in the Biology Department of the Université Libanaise (Hadeth-Beyrouth). The late Y. Palmoni was the first person to systematically collect adult caddisflies in the Levant (in Israel). Most of the specimens studied for this book were collected either by myself, or by Dr A. Gasith (Tel Aviv), members of the already mentioned I.E.S. (Jerusalem), Dr A. Dia (Beirut) and Dr Z. Moubayed (Montpellier).

1

Notes concerning the illustrations and text

In many plates, the various figures are to different scales. In several figures of genitalia some details are omitted to avoid unnecessary complication. Only exceptionally are such situations mentioned in the legend.

If not otherwise specified, the drawings are original and were prepared from Levantine specimens.

The numbers in parenthesis which appear after the locations refer to the geographical areas on the map of Israel and Sinai, at the end of the book. The spelling of names of localities in Israel is according to the maps published by the Survey of Israel.

INTRODUCTION

MORPHOLOGY

It is possible, in some instances, that the entomologist catching adult Trichoptera will experience some difficulty, in distinguishing them rapidly from small Lepidoptera. Nevertheless, two characters of the latter order are never, or practically never present in Trichoptera: the parts of the maxillae known as galea are very often strongly modified to form a longer or shorter, coiled sucking tube; and there is always a covering of scales with metallic gleam on the wings and other parts of the body (there are a few non-Levantine Trichoptera with some scales, e.g., on the forewings). Other confusions — for instance with small Neuroptera — will be rapidly avoided by examination with a magnifying glass.

Caddisflies are always setose, often very setose, insects. On their head, besides two large, globose, composite eyes, three ocelli are often found on the vertex (Fig. 1), located in the angles of a triangle (they may be concealed by setae, but are generally easily seen). The antennae are varied morphologically in the different families, the 1st segment (scapus) always being distinct from the following segments which are more or less homonomous (scapus generally longer and thicker, sometimes strongly modified). Very important in the identification of the families are the maxillary palpi, always well developed and distinctly longer than the (generally 3-segmented) labial palpi; they are 5-segmented in most families represented in the Levant. The relative length of the segments is important, as well as the fact that, whereas in some families the 5th segment is normal, rigid, and not very long (Fig. 2), in others (Fig. 3) it is very long, clearly flexible, and apparently formed of a large number of small secondary joints; in some Leptoceridae (which, anyway, will be rapidly distinguished by their very long and slender antennae) the situation is somewhat intermediate: 5th segment flexible (as may be the distal part of segment 4) but not annulate; there are less than 5 segments in the males of Limnephilidae (where they are normal), Brachycentridae (slightly modified), and Sericostomatidae (strongly modified); there are 4 segments in Phryganeidae, a family not yet known from the Levant.

The morphology of the dorsal parts of the head and thorax (shape of various parts separated by sutures, smaller or larger setose warts) is complex and can offer excellent characters for classification at all levels. Unfortunately, I was unable to take these characters into account in the present volume. A well illustrated study of these structures in all the Levantine species would certainly be worth while. In difficult cases of correct association of the two sexes, examination of these structures can give excellent results, because they are mostly identical in males and females.

The legs, long and slender (tibiae and tarsi of intermediate legs sometimes dilated in female, Fig. 5), have 5-segmented tarsi. Besides short, rigid spines, there are always

3

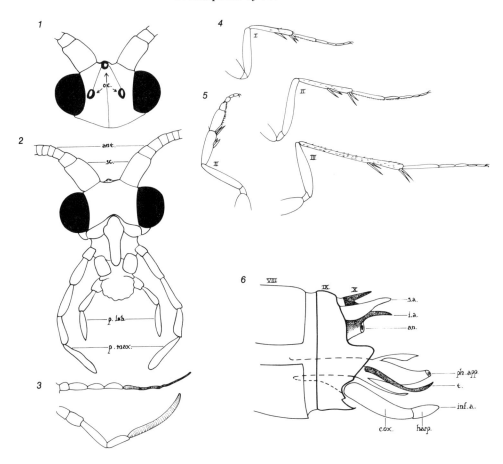

Figs. 1–6: Diagrammatic representation of some morphological aspects
of adult Trichoptera
1. head, dorsal view; 2. head and appendages, frontal view;
3. maxillary palpi with flagellar (microsegmented, flexible) last segment;
4. legs of the three pairs, with their tibial spurs (formula: 3, 4, 4);
5. intermediate leg (in females of some Trichoptera) with dilated tibia and tarsus;
6. strongly generalized male genitalia

an. – anus; ant. – antenna; cox. – coxopodite; harp. – harpago;
i.a. – intermediate appendage; inf. a. – inferior appendage; oc. – ocelli;
p. lab. – labial palpi; p. max. – maxillary palpi;
ph. app. – phallic apparatus; s.a. – superior appendage;
sc. – antennal scapus; t – titillator
Slightly modified from various sources

some movable spurs on the tibiae (Figs. 4–5), whose number and arrangement is constant in each species and often very characteristic at generic or familial level. These spurs are practically always easily distinguished from the normal spines, even by the beginner, because they are longer and otherwise coloured (sometimes, the apical spurs on the foretibia can be very small); they are present either apically on the tibiae (apical spurs) or more or less in the middle (subapical spurs). Very different spur formulae are found, the highest number being 3, 4, 4 (*Rhyacophila, Polycentropus, Pseudoneureclipsis, Ecnomus*), the lowest being 0, 2, 2 (*Oecetis, Setodes*). In the first case this means that there are 3 spurs on the foretibia (one apical pair, one subapical spur) and 4 spurs on the intermediate and hind tibiae (one apical and one subapical pair); in the second case, there is no spur on the foretibia, and there is one pair of apical spurs on the other two tibiae. If there are only 1 or 2 spurs on a tibia, they will be apical; should a 3rd or also a 4th be present, they will be subapical. It is important to correctly count the spurs. It should be taken into account that a spur may be broken; even then, its alveolus will be observed.

At rest, the forewings of caddisflies characteristically always form a "roof". Shape, colour pattern and vestiture of the two pairs of wings are extremely diverse in Trichoptera, and their venation offers extremely important characters at supraspecific levels. In order to avoid a detailed description, I have marked on the diagrammatic and generalized Fig. 7 all the terminology used throughout this volume. Only a few remarks are needed here. In most recent publications, the vein following after that forming apical fork 5 in both wings is called the 2nd cubital vein (Cu2); I am conservative in naming it first anal vein (A1), the branches of f5 being for me Cu1 and Cu2 instead of Cu1a and Cu1b. The apical forks are, in my opinion, extremely important and useful characters, and I completely disagree with authors who neglect them in their descriptions. In families like Hydroptilidae, Leptoceridae and Beraeidae, the venation can be drastically simplified, anomalous and difficult to interpret; I have illustrated it in all genera in order to lessen difficulties. There is sometimes some sexual dimorphism in the wings. Finally, slight anomalies in the venation are not a very exceptional phenomenon.

There are sometimes various formations (filaments, rigid projections) on the pregenital abdominal segments. Ventrally in abdominal segment V, in both sexes in many caddisflies, there is a pair of glands almost certainly involved in pheromone production, and externally indicated by some formations; I have mentioned and illustrated these organs only when they are very conspicuous (*Agapetus*).

The genitalia are of utmost importance in the study of Trichoptera: the main lines of their architectonics supply excellent characters for the description of genera, their details being the main source of characters for specific classification. It is practically impossible to give a concise description of male or female genitalia, the variations being of tremendous amplitude. But, in their description for every species I have done my best to help the reader to recognize without difficulty the respective structures on the illustration. I have been pragmatic and have tried to adopt a simple and uniform terminology, based more on the position of the appendages than on their origin.

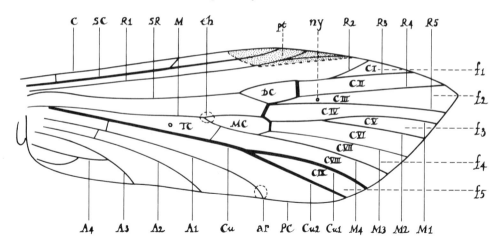

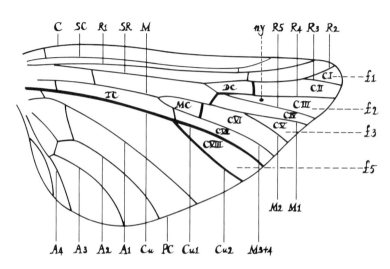

Fig. 7: Strongly generalized fore- and hindwing of a caddisfly
A1–A4 – anal veins; ar – arculus; C – costal border;
CI–CIX – apical cells; Cu, Cu1, Cu2 – cubitus;
DC – discoidal cell; f1–f5 – apical forks;
M – media; M1–M4 – branches of M; MC – median cell;
ny – nygma; PC – postcostal border; pt – pterostygma;
R1 – radius; R2–R5 – branches of SR; SC – subcosta;
SR – sector radii; TC – thyridial cell; th – thyridium

I give here a diagrammatic and generalized drawing of the male genitalia (Fig. 6). In the male, the last non-modified abdominal segment is often segment VIII, but even this segment is sometimes slightly modified. Segment IX is represented by a sclerotized capsule; on the distal margin of its dorsal part (described as tergite even if there is no clear limit between tergite and sternite) the superior appendages are rooted; connected to its ventral part (sternite) are the inferior appendages, sometimes also called gonopods, often bisegmented (coxopodite and harpago). In most cases it is easy to distinguish the parts formed by the Xth (and last) segment, which has a dorsal position in the genitalia, being often represented by a central body with lateral appendages called "intermediate appendages"; it is here that the anal opening is located. Below segment X there is the phallic complex or apparatus (sometimes simply called phallus) consisting — when it is complete, from root to tip — of a sclerotized phallotheca, a membranous endotheca, an aedeagus (inside which the ductus ejaculatorius leads to the genital opening), and one or two titillators.

In the female, interpretation of the genital segments is sometimes merely conjectural. Generally, segments VIII, IX, and X are involved; I was able to recognize a segment XI only in *Rhyacophila* (with hesitation!), in Glossosomatidae, and in some Hydroptilidae. Some sclerotized internal structures are systematically important; they will be seen, of course, only on abdomina correctly cleared in KOH.

SYSTEMATICS AND PHYLOGENY

To give the reader an idea of the main divisions of the Order Trichoptera with a complete list of families, the most recent and complete phyletic tree of the order is reproduced here (Fig. 8) from Weaver & Morse (1986). In this figure, families represented in the Levant are marked with a circle.

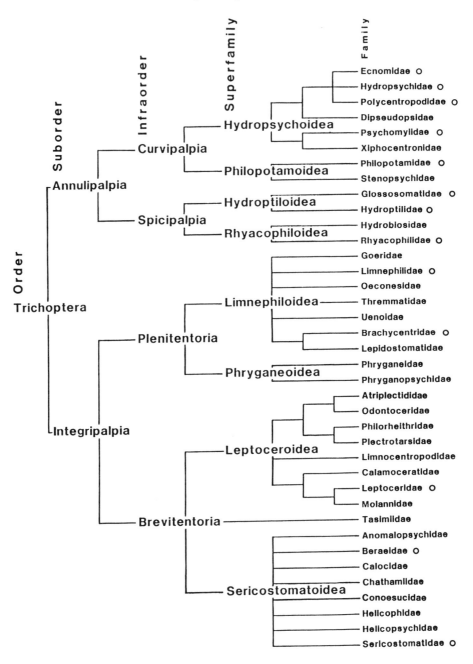

Fig. 8: Phylogeny of Trichoptera with complete list of families
(from J.S. Weaver & J.C. Morse, 1986)
Families represented in the Levant are marked by a small circle

8

BIOLOGY AND ECOLOGY

Trichoptera are, in their vast majority, amphibiotic (or hemihydrobiont: Berczik, 1973) insects. Larvae and pupae develop in the most diverse freshwater habitats. In all species they are in some manner involved in building activities — an extremely interesting and fairly unique phenomenon. They occupy all possible kinds of ecological niches and belong to all "feeding guilds": predators, herbivores, piercers, scrapers, shredders, gatherers, filterers. Adults generally live in the vicinity of the same habitats, and they will be found on, or in, all kinds of vegetation along banks or shores (some plants are probably more attractive than others). They are, generally, very secretive insects.

Most species are poor flyers, being found only in the close vicinity of water. The best flyers are probably species of *Hydropsyche* and especially of Limnephilidae (*Limnephilus, Stenophylax*) which are able to fly rather far from their breeding places. Several distinct types of flight are distinguished, e.g., upstream directed flight along water courses; flight from the banks to the stream and vice versa; swarming flight related to search for partners; flight related to egg-laying; in some instances (Hydropsychidae, Leptoceridae) there are conspicuous "dances" of males in search of females.

Many caddisflies are nocturnal; they are readily attracted by artificial light (sometimes with notable differences between sexes), and are also active early in the morning and at twilight — the best times to catch them by net. Less (much less) attracted by light are, e.g., crenobiont species or species from madicolous habitats (see below). There are also species very active during the day, even in full sunlight (like some Leptoceridae).

Phenological observations will be given in this volume for practically every species, but for many of them the existing information is very scanty. It is not surprising that for some species a longer flight period (including the winter months) is observed in the Levant than, say, in central or northern Europe. There are extremely few exact data concerning the number of generations per year in various species in the Levant. All this is a vast field for future observations.

In many species the imaginal emergence is nocturnal, in many others diurnal, and every species has its own rhythm. With the remarkable exception of species with long aestival diapause or dormancy (Limnephilidae), females emerge with fully developed gonads, and activities related to reproduction start rapidly.

Copulation follows almost immediately after emergence; in some caddisflies it is preceded by the swarmings (dances) already mentioned. In most cases, copulation takes place on plants; partners generally copulate with heads in opposite directions; it lasts from a few minutes to a few hours. Repeated copulation is known in many cases, with egg laying occurring in between. Normally, there is an interval of a few hours between copulation and egg laying.

For oviposition, females of many species sink into the water, eggs being fastened onto different substrata, but there are also other kinds of oviposition (on the water surface; on substrata above the water surface, along banks or shores). There are three main types of oviposition: eggs are individually deposited (e.g., in *Rhyacophila*); eggs are produced in strings, several strings forming a thin plate well fastened to the substrate (Philopotamidae, Polycentropodidae, Ecnomidae, Psychomyiidae, Hydropsychidae); egg masses are produced, surrounded by a substance which swells in contact with water, becoming voluminous, mostly spherical or hemispherical translucent masses (Brachycentridae, Limnephilidae, Leptoceridae, Sericostomatidae, Beraeidae). In these latter families (Limnephilidae excepted) females carry the egg masses with them, in flight, before oviposition.

Adult caddisflies drink mainly water, but there are observations on flower-visiting species which certainly drink nectar, and on insects sucking sweet fruit juices.

Caddisflies are generally more or less closely associated with certain aquatic biotopes, the truly ubiquitous species being exceptions. Moreover, they are present, sometimes forming large or huge populations and associations, in practically every kind of freshwater habitat, except heavily polluted water. An attempt will be made here to group the species together according to their different habitats. Of course, this will only reflect the situation in the Levant: we have to take into account the phenomenon of ecological vicariance (Botosaneanu, 1960). For problems and terminology concerning longitudinal zonation of running water, see Illies & Botosaneanu (1963), Botosaneanu (1979), Dia (1983), and Moubayed (1986).

Hygropetric, or madicolous, habitats (water — especially of springs, or near larger or smaller waterfalls — flowing in a very thin layer on more or less vertical rocky surfaces): *Stactobia caspersi, S. margalitana*(?), *S. aoualina, S. pacatoria, Hydroptila viganoi*(?).

Crenal (springs and spring brooks): *Agapetus caucasicus, Agraylea (Allotrichia) teldanica, Hydroptila viganoi, Oxyethira falcata, O. assia, Hydropsyche longindex, Tinodes israelicus, T. negevianus, T. kadiellus, Apatania cypria, Ernodes saltans.*

Crenal and Rhithral (springs and streams): *Rhyacophila fasciata, Agraylea (Allotrichia) vilnensis orientalis*(?), *Hydroptila sparsa, H. angustata, H. phoeniciae, H. aegyptia, H. vectis, Oxyethira falcata, Orthotrichia moselyi, Ithytrichia lamellaris, Hydropsyche instabilis, Polycentropus flavomaculatus hebraeus, P. baroukus, Psychomyia pusilla, Lype reducta, Tinodes caputaquae, T. tohmei, Limnephilus lunatus, Adicella syriaca, Oecetis terraesanctae.*

Rhithral (different zones of streams, including the largest streams or "mountain rivers"): *Rhyacophila nubila, Agapetus orontes*(?), *Glossosoma hazbanicum, G. capitatum*(?), *Hydroptila mendli levanti, H. palaestinae*(?), *H. libanica, H. fonsorontina, Oxyethira delcourti, Wormaldia subnigra, Cheumatopsyche capitella, Hydropsyche jordanensis, H. pellucidula, H. theodoriana, Pseudoneureclipsis palmonii, Ecnomus gedrosicus, Brachycentrus (Oligoplectrum) maculatus, Setodes viridis huliothicus, S. kugleri, Sericostoma flavicorne.*

Crenal and Rhithral of desert zones: *Hydroptila adana, H. hirra, Ithytrichia dovporiana, Chimarra lejea, Tinodes negevianus, Mesophylax aspersus, Setodes alala.*

Epipotamal (lower courses of Rivers Jordan and/or Litani): *Hydroptila angustata, Orthotrichia melitta*(?), *Cheumatopsyche capitella, Hydropsyche pellucidula, H. janstockiana, H. batavorum* (probably extinct), *Psychomyia pusilla, Ceraclea (Athripsodina) litania, Ylodes internus* (probably extinct population), *Setodes kugleri.*

Lake Kinneret: *Orthotrichia moselyi, Pseudoneureclipsis palmonii, Ecnomus galilaeus, E. gedrosicus.*

H̲ula marshes: *Tricholeiochiton fagesii, Orthotrichia costalis, Ylodes reuteri* (all probably extinct populations).

Water courses, or pools and small lakes, drying up (or with very warm water) in summer (adults with long estival dormancy, either in caves, or in other dark hollows and cracks): *Stenophylax vibex, S. permistus, S. tauricus, S. coiffaiti, S. malaspina, S. lindbergi, Mesophylax aspersus.*

GEOGRAPHICAL DISTRIBUTION

The Levantine Province includes, in my opinion, the Sinai, Israel, Jordan, Lebanon, and Syria west of the Euphrates River. Some parts of the Province are fairly well explored with regard to the caddisfly fauna (Sinai, most of Israel, important parts of Lebanon), whereas very little is known in this respect about other parts (the Judean Hills, Samaria, Jordan, the far north of Lebanon, Syria). The number of known species and subspecies is relatively low (about 80), but some 30 of them are presently known only from the Levant, indicating a relatively high proportion of potentially endemic elements (two somewhat doubtful cases are included here, viz. *Hydroptila mendli levanti* and *Hydroptila palaestinae*; see the parts devoted to these two taxa). Careful collecting in the extreme north of Lebanon and in north-west Syria will certainly add some more species to the list, while further studies in Jordan, the Judean Hills and Samaria are not expected to substantially modify the caddisfly list; the knowledge of the Syrian fauna as a whole is very fragmentary. Eight families whose presence in the Levant could be expected are for the time being not recorded (Phryganeidae, Goeridae, Thremmatidae, Lepidostomatidae, Helicopsychidae, Odontoceridae, Calamoceratidae, Molannidae); nine other families are poorly represented (Rhyacophilidae, Glossosomatidae, Philopotamidae, Polycentropodidae, Ecnomidae, Brachycentridae, Limnephilidae, Beraeidae, Sericostomatidae). Best represented are the Hydroptilidae (especially *Hydroptila*), Psychomyiidae (especially *Tinodes*), Hydropsychidae, and Leptoceridae. The Levantine Province has a very marked "cross-road" character, clearly reflected in its caddisfly fauna.

Recent research shows that at least four zones, each with a distinct trichopteran fauna, can be distinguished within the limits of the Levant. Three of them clearly correspond to different hydrographic basins rather than to orographic systems (zone I: catchments of the Orontes and of the Litani; zone II: coastal basins of Lebanon; zone III: catchment of Jordan River, including Lake Kinneret); the fourth zone includes the desert areas of the province, its northern boundary being the northern limit of the Dead Sea Depression. Lists of the most characteristic taxa presently known from these four zones (and also from the Judean Hills; see Appendix) are given here. Those taxa which are exclusively known from one or several zones of the Levant, being the potential endemics of the Province, are printed in bold face; italics are used for other taxa (i.e., not potentially endemic), whose distribution in the Levant is apparently restricted to one or two of the above-mentioned zones. In a series of cases, such restricted distribution patterns seem to be more or less relevant, and, e.g., several cases of interesting replacement (vicariance) of species between zones I, II, and III are listed in Moubayed & Botosaneanu (1985).

The fauna of Lebanon and the northern half of Israel is dominated by elements of western Palaearctic distribution and affinities, mainly with an eastern Mediterranean and Balkan tint, and sometimes with a slight Anatolian, Iranian, Central Asiatic, or North African tint. There is nothing remarkable about all this, and examples are not needed. This is particularly true for Lebanon, where practically the entire caddis fauna has this character. Somewhat surprising is the presence, in considerable populations and at many localities in the coastal basins of Lebanon, of some European species in their most typical morphological aspect (*Wormaldia subnigra*, *Hydropsyche pellucidula*) which disappear further south, in Israel, or become distinctly less abundantly represented there. Worthy of mention here are some interesting elements shared by Lebanon (some of them by northern Israel too) with some of the eastern Mediterranean islands: *Apatania cypria* (Cyprus), *Tinodes kadiellus* (Rhodes, Cyprus), *Oxyethira delcourti* (Cyprus, Crete, Aegean Islands), *Orthotrichia melitta* (Lesbos). Three *Tinodes* species not found outside the Levant (*israelicus*, *negevianus*, *caputaquae*) belong to a species-group best represented in the eastern Mediterranean. There are also several Caucasian, Anatolian, or northern Iranian elements whose southernmost limits of distribution are in Lebanon (*Glossosoma capitatum*, *Agraylea vilnensis orientalis*, *Ernodes saltans*, *Sericostoma flavicorne*) or even reach Israel (*Agapetus caucasicus*, *Stenophylax lindbergi*, *S. malaspina*). The Caucasian and insular species mentioned above inhabit rather high mountains, whereas water courses at lower altitudes have a predominantly eastern Mediterranean fauna, and also feature some of the very few elements of possible Oriental affinities, such as *Hydroptila libanica* and *H. fonsorontina*.

In northern Israel — as far as the northern limit of the Dead Sea Depression — the fauna, too, is clearly dominated by western Palaearctic elements, with the several "tints" already mentioned, the springs of the Jordan River and especially those at Tel Dan probably being the gathering place for the largest number of species. It is noteworthy that some typical European species abundantly represented in Lebanon (see above) are absent from Israel; they were possibly eliminated from Galilee and from the coastal plain by human activities. Moreover, several Caucasian and insular species present in Lebanon do not reach Israel: *Apatania cypria, Ernodes saltans, Sericostoma flavicorne* (and this holds for other species too, like *Rhyacophila fasciata* or *Brachycentrus (Oligoplectrum) maculatus*). Several species with western Palaearctic distribution and/or affinities have their southernmost known localities in Israel: *Rhyacophila nubila, Agapetus caucasicus, Glossosoma hazbanicum, Agraylea teldanica, Hydroptila sparsa, H. libanica, Ithytrichia lamellaris, Lype reducta, Tinodes caputaquae, T. israelicus, T. kadiellus, Adicella syriaca, Setodes viridis huliothicus.* Perhaps the most interesting constituent of the fauna of Israel and adjacent areas is a small group of species which, in my opinion, have an Oriental origin. These apparently penetrated into south-west Arabia and subsequently radiated, especially northwards to the Sinai, the Negev, and the Dead Sea Depression. These species could be characterized as "eremic" elements among the caddisflies, inhabiting essentially relict, isolated freshwater springs and streams in xeric environments. This category includes the following species: *Hydroptila adana, H. hirra, Chimarra lejea, Setodes alala.* Some of these species were very expansive: *Hydroptila hirra* and *Chimarra lejea* have also invaded parts of the Ethiopian Region, and *H. hirra* progressed northwards to the Jordan Valley (the same may be true for *H. palaestinae*, if this species is indeed present in south-western Arabia, as mentioned in the literature). Finally, it should be added that quite a few species with western Palaearctic distribution and/or affinities accompany this complex of "eremics" in the deserts of the Levantine Province: *Hydroptila angustata, H. vectis,* two species of *Stactobia, Psychomyia pusilla, Tinodes negevianus, Hydropsyche instabilis, Mesophylax aspersus.*

A few special cases of Oriental affinities should be mentioned here. The first is that of *Stactobia pacatoria*, with distribution restricted to Lebanon, but clearly an offshoot of a lineage of the same origin as that of the species mentioned above. The second is that of *Hydroptila palaestinae*, possibly belonging to the same distributional category, but, in any event, belonging — together with its sister-species *H. libanica* and *H. fonsorontina* — to the same circle of affinities. Finally, there is the case of *Pseudoneureclipsis palmonii*, another species with Oriental affinities.

Special mention should be made of the rather surprising fact that two Levantine species (*Oxyethira assia*, from the upper course of the Orontes, Lebanon, and *Ithytrichia dovporiana*, from Naḥal 'Arugot, Dead Sea Depression, Israel) have distinct affinities to northern species, Scandinavian in the first case, Nearctic in the second, i.e., two cases of "Boreo-Levantine disjunction"?

It is a well-known fact that the Levantine fauna in various groups includes an Ethiopian component which is sometimes important. However, only very few species of Trichoptera can be considered as possibly having an Ethiopian origin: *Hydroptila aegyptia, Orthotrichia moselyi, O. melitta, O. costalis, Chimarra lejea, Ecnomus galilaeus.* The case of *Oecetis terraesanctae* is even less clear (it might be of either Ethiopian or Oriental origin). It is noteworthy that two of the above-mentioned species belong to the very poor caddisfly fauna of Lake Kinneret.

A reconstruction of the main historical trends in the establishment of the present Levantine trichopteran fauna was attempted by Botosaneanu (1973). This process can be summarized as follows: migration northwards along the Great Rift Valley, after the separation of Arabia and Africa, in the Upper Pliocene, of several Oriental lineages that had established themselves during the Upper Miocene and Lower Pliocene in the south-west of the Arabian Peninsula. This northward progression was stopped in post-Würmian times by the southward progression of many Palaearctic lineages, *several* such southward migration waves having probably occurred during the Pleistocene, in connection with the complex history of the formation of the Jordan and of the Orontes basins. The Levantine Province obviously did not serve as an important refuge for western Palaearctic faunas during the Pleistocene glaciations, nor was it an important recolonization centre, quite unlike several other Provinces around the Mediterranean.

List of some of the more characteristic species presently known from the different zones of the Levantine Province
(**bold face** — potentially endemic elements of the Levant; *italics* — elements with wider distribution, but apparently restricted, in the Levant, to one or two zones).

Zone I: catchments of the Orontes and of the Litani. *Rhyacophila fasciata* Hagen, **Agapetus orontes** Malicky, **Glossosoma hazbanicum** Botosaneanu & Gasith, *G. capitatum* Martynov (occurrence not quite certain), *Hydroptila sparsa* Curtis, **H. fonsorontina** Botosaneanu & Moubayed, **Oxyethira (O.) assia** Botosaneanu & Moubayed, **Orthotrichia moselyi** Tjeder, **Hydropsyche janstockiana** Botosaneanu (occurrence not quite certain), *Polycentropus flavomaculatus hebraeus* Botosaneanu & Gasith, **Tinodes caputaquae** Botosaneanu & Gasith, *Brachycentrus (Oligoplectrum) maculatus* (Fourcroy), *Apatania cypria* Tjeder, *Limnephilus lunatus* Curtis, **Ceraclea (A.) litania** Botosaneanu & Dia, *Sericostoma flavicorne* Schneider.

Zone II: coastal basins of Lebanon. *Rhyacophila fasciata* Hagen, **Glossosoma hazbanicum** Botosaneanu & Gasith, **Stactobia aoualina** Botosaneanu & Dia, **S. pacatoria** Dia & Botosaneanu, *Agraylea (Allotrichia) teldanica* (Botosaneanu), *A. (A.) vilnensis orientalis* Botosaneanu, **Hydroptila phoeniciae** Botosaneanu & Dia, **H. mendli levanti** Botosaneanu, **H. libanica** Botosaneanu & Dia, *Oxyethira (O.) delcourti* Jacquemart, *Orthotrichia melitta* Malicky, *Wormaldia subnigra* McLachlan, **Hydropsyche longindex** Botosaneanu & Moubayed, *H. pellucidula* (Curtis), **H. theodoriana** Botosaneanu, **Polycentropus baroukus** Botosaneanu & Dia, **Tinodes caputaquae** Botosaneanu & Gasith, **T. israelicus** Botosaneanu & Gasith, **T. negevianus** Botosaneanu & Gasith, **T. kadiellus** Botosaneanu & Gasith, **T. tohmei** Botosaneanu & Dia, *Apatania cypria* Tjeder, *Stenophylax vibex* (Curtis), *S. coiffaiti* (Décamps), **Setodes viridis huliothicus** Botosaneanu & Gasith, **S. kugleri** Botosaneanu & Gasith, *Ernodes saltans* Martynov.

Zone III: catchment of Jordan River. **Glossosoma hazbanicum** Botosaneanu & Gasith, *Agraylea (Allotrichia) teldanica* (Botosaneanu), *Hydroptila sparsa* Curtis, **H. palaestinae** Botosaneanu & Gasith, **H. libanica** Botosaneanu & Dia, *H. hirra* Mosely, **Orthotrichia moselyi** Tjeder, *O. costalis* (Curtis), **Hydropsyche jordanensis** Tjeder, *H. pellucidula* (Curtis), **H. theodoriana** Botosaneanu, **H. janstockiana** Botosaneanu, **H. batavorum** Botosaneanu, *Polycentropus flavomaculatus hebraeus* Botosaneanu & Gasith, **Pseudoneureclipsis palmonii** Flint, **Ecnomus galilaeus** Tjeder, *E. gedrosicus* Schmid, **Tinodes caputaquae** Botosaneanu & Gasith, **T. israelicus** Botosaneanu & Gasith, **T. negevianus** Botosaneanu & Gasith, *T. kadiellus* Botosaneanu & Gasith, **Limnephilus**

turanus hermonianus Botosaneanu (only known from Mount Hermon), *Ylodes internus* (McLachlan), *Y. reuteri* (McLachlan), **Oecetis terraesanctae** (Botosaneanu & Gasith), **Setodes viridis huliothicus** Botosaneanu & Gasith, **S. kugleri** Botosaneanu & Gasith.

Zone IV: the desert areas. **Stactobia margalitana** Botosaneanu, *Hydroptila adana* Mosely, *H. hirra* Mosely, **Orthotrichia moselyi** Tjeder, **Ithytrichia dovporiana** Botosaneanu, *Chimarra lejea* Mosely, **Tinodes negevianus** Botosaneanu & Gasith, *Setodes alala* Mosely.

Appendix: the Judean Hills. **Hydroptila viganoi** Botosaneanu, **Tinodes negevianus** Botosaneanu & Gasith.

COLLECTING, CONSERVING AND PREPARING ADULT TRICHOPTERA

Adult caddisflies are collected either with a soft fabric net by sweeping the vegetation along water courses and on shores of stagnant water bodies, or by directly picking or sweeping the insects sighted on vegetation or on objects near water, and finally with every kind of artificial light (an ultra-violet lamp or a Coleman lamp would both give excellent results).

Dry and pinned insects will form a fine collection but one not easy to study. Insects kept in 70% alcohol will provide, perhaps, a less fine collection, but are much more easily studied. The ideal formula would be two parallel collections.

In order to study the wing venation, it will be necessary rather often to remove, at least partially, the vestiture of setae from the wings. This operation is by no means easy and should be performed with an extremely fine brush, with one mounted hair, or with the finest entomological pin; it will be more easily performed in fluid. Various lightings should be tried, on all kinds of background, for examining the venation in species where it is indistinct. The beginner will have to carefully clear the abdomen of almost every specimen in a (cold or warm) KOH solution in order to study the genitalia (almost all the genitalia drawings in this volume were made from cleared abdomina). More experience will allow, in some cases, the satisfactory examination of non-cleared specimens, or the clearing in KOH of only one specimen in a series; but in very small insects, like Hydroptilidae, this operation is almost unavoidable. Keeping the abdomina too long in KOH can result in a disaster. Cleared abdomina should be well washed (internally and externally) in alcohol, examined e.g. in glycerin, kept in small tubes with glycerin or with 70% alcohol, accompanying the respective pinned or alcohol preserved specimen. Fine microslides can be prepared (in Canada balsam, Euparal, etc).

SYSTEMATIC PART

Key to the Families of Trichoptera in the Levant[1]

1. Minute to small[2], very setose insects, forewings also with thickened, erect setae, fringes extremely long — especially in hindwings. Antennae shorter than forewings (shorter in ♀ than in ♂), generally with very short, rounded segments. Wings often very narrow, sharply ending, with simplified venation, discoidal cell absent (open) in hindwings. Maxillary palpi (♂ , ♀) with 5 segments, last segment normal. **Hydroptilidae**

   – Small to large insects (not minute), without thickened, erect setae on forewings, fringes of wings not extremely long (those of hindwings not longer than greatest breadth of wing). Antennae as long as, or longer than, forewing   2

2. Maxillary palpi (in both sexes, or only in ♀) with 5 segments   5

   – Maxillary palpi in ♂ with less than 5 segments   3

3. Ocelli present. No more than 1 spur on foreleg.   **Limnephilidae** ♂♂

   – No ocelli. 2 spurs on forelegs   4

4. Spurs 2, 2, 4. Last segment of maxillary palpi enormously developed, like a "mask" adpressed to the head.   **Sericostomatidae** ♂♂

   – Spurs 2, 2, 2 (or 2, 3, 3). Maxillary palpi adpressed to head, very setose, not strongly modified.   **Brachycentridae** ♂♂

5. Terminal segment of maxillary palpi secondarily divided into many minute joints, flexible, longest of all segments. Median cell in forewing closed   6

   – Terminal segment of maxillary palpi not secondarily divided into minute joints, not very long, generally rigid[3]   10

6. Ocelli present.   **Philopotamidae**

   – Ocelli absent   7

7. Antennae strong, with short segments. Spurs 3, 4, 4 or 2, 4, 4. Inferior appendages of ♂ unisegmented (not divided in coxopodite and harpago)   8

   – Antennae more slender or even very slender, segments more elongate. Spurs 2, 4, 4. Inferior appendages of ♂ bisegmented (with distinct coxopodite and harpago)   9

8. 2nd segment of maxillary palpus only as long as 1st, much shorter than 3rd. Discoidal cell in forewing more or less elongate. R1 in forewing not forked at end.   **Polycentropodidae**

   – 2nd segment of maxillary palpus longer than 1st, as long as or longer than 3rd. Discoidal cell in forewing short. R1 in forewing forked at end.   **Ecnomidae**

9. Wings slender, hindwings pointed at apex and even more slender than forewings. Apical fork 1 absent in both wings. ♂ with well-developed superior appendages. ♀ with ovipositor.   **Psychomyiidae**

---

[1] To be used with caution; check with family diagnoses.

[2] But there are other caddisflies (*Agapetus, Ecnomus, Psychomyia, Lype* ...) as small as, or smaller than, larger Hydroptilidae.

[3] Flexible only in some Leptoceridae (but not annulate!).

17

–	Hindwings not pointed, at least as broad as forewings. Apical fork 1 present at least in forewing. ♂ without superior appendages. ♀ with short genitalia (no ovipositor). **Hydropsychidae**	
10.	Ocelli present	11
–	Ocelli absent	13
11.	Forewing with all apical forks (1–5). Spurs 3, 4, 4 or 2, 4, 4. First two segments of maxillary palpi very short	12
–	Forewing (and hindwing) with forks 1, 2, 3, 5. No more than one spur on foreleg and 3 spurs on intermediate legs. 2nd segment of 5-segmented maxillary palpi longer than 1st. **Limnephilidae** ♀♀	
12.	Spurs 3, 4, 4. Discoidal cell in both wings open. **Rhyacophilidae**	
–	Spurs 2, 4, 4. Discoidal cell in forewing closed. **Glossosomatidae**	
13.	Discoidal and median cells in both wings absent. Spurs 2, 2, 4 **Beraeidae**	
–	Discoidal cell in forewing closed	14
14.	Antennae very slender and long (much longer than forewings), with long segments. No apical fork 2 in forewings[4]. Forewings elongate and slender (but hindwings sometimes broad!). **Leptoceridae**	
–	Antennae at most as long as forewings, not very slender. Fork 2 in forewings present. Forewings moderately broad	15
15.	Four spurs on hindlegs (spur formula mostly 2, 2, 4). **Sericostomatidae** ♀♀	
–	Only 2 or 3 spurs on hindlegs (spur formula: 2, 2, 2 or 2, 3, 3). **Brachycentridae** ♀♀	

Family RHYACOPHILIDAE Stephens, 1836
Ill. Br. Ent., 6:148, 154

Type Genus: *Rhyacophila* Pictet, 1834.

This diagnosis is valid for the subfamily represented in the Levant: Rhyacophilinae.

Head with ocelli. Maxillary palpus with 5 segments in both sexes, segments 1 and 2 very short, last segment simple and apically pointed (like last segment of labial palpus). Antennae slender, at most as long as forewings, scapus stout, short, not modified. Spurs long: 3, 4, 4. Tibiae and tarsi of intermediate legs in the female not dilated. Wings regularly elliptical, similar in the two pairs and in both sexes, with short and not very dense setae; venation very complete, with forks 1–5 in forewing, and 1, 2, 3, and 5 in hindwing; discoidal (and median) cell in both wings open. Male genitalia complex, with many different appendages, gonopods large, bisegmented, parts belonging to abdominal segment XI sometimes recognized. Female with well-developed ovipositor.

One genus represented in the Levant: *Rhyacophila* Pictet.

---

[4] In *Ylodes* there is a false f 2.

18

Genus RHYACOPHILA F.J. Pictet, 1834
Rech. Phryg., pp. 23, 181

Type Species: *Rhyacophila vulgaris* Pictet, 1834; subsequent designation by Ross, 1944.

Diagnosis: Rather large insects. Forewings greyish-yellow or brownish with a distinct reticulation determined by areas of paler or darker setae (females darker than males), setae on forewings short, not very dense, leaving venation clearly distinct (Fig. 9). Antennae shorter than forewings, slender, with very short basal segment. Ocelli present, large. Maxillary palpus with two very short, rather thick basal segments, followed by three slender and much longer segments (4th segment shorter than 3rd and 5th, this last one ending in sharp point). Spur formula in both sexes 3, 4, 4, spurs long, the pairs nearly equal. Legs long and slender, tibiae and tarsi of 2nd pair in the female not dilated. Fore- and hindwings without sexual dimorphism, rather similar in shape and venation, elongated (elliptical) with more or less parabolic apex, hindwings somewhat shorter but not broader than forewings, and not folded. Pterostigma distinct. Discoidal cell open in both wings; in forewing apical forks 1, 2, 3, 4, 5; in hindwings apical forks 1, 2, 3, 5. On abdominal sternite VII, a small median tooth in both sexes. Male genitalia with following characters: segment IX well-developed laterally and dorsally, much less so ventrally; medio-dorsally produced in an apical lobe, differently shaped in the species; beneath this lobe, a horizontal plate formed by the coalescent preanal appendages; perpendicular to this plate is a massive sclerotized vertical formation known as "body of segment X"; in its lower part, a pair of heavily sclerotized, dark, baculiform processes having a common root, and known as "anal sclerites"; the inferior appendages (gonopods) are the most conspicuous part of the genitalia, they are bisegmented, 1st segment much longer than 2nd segment (harpago) which is variously shaped; in the copulatory structures, the distal, sclerotized parts are the most easily distinguished; the central (impair) structure is the aedeagus, having a dorsal branch (aedeagus proper) and a ventral one (the "ventral lobe"); laterally to the aedeagus there is a pair of mobile parameres. In the female genitalia it is segment VIII which gives the best distinctive characters, segments IX–X (or XI) being normally telescoped within it; basally on segment VIII there is a sclerotized "collar" surrounding a longer, less chitinous part; the "collar" is distinctly split dorsally and ventrally.

General Remarks: *Rhyacophila* is an extremely large genus with about 500 known species inhabiting most of the Palaearctic, Oriental, and Nearctic regions. With only two known species, the genus is extremely poorly represented in the Levant; the discovery of some additional species (possibly Caucasian, or Anatolian) in the province (in northern Lebanon or in adjacent zones of Syria) is neither impossible, nor very probable. Both species presently known from the province belong to one large species-group ("group *vulgaris*": Schmid, 1970) and are rather closely related.

Rhyacophila is considered as being the most primitive genus of Trichoptera. Its species are always inhabitants of running water, and they are in their overwhelming

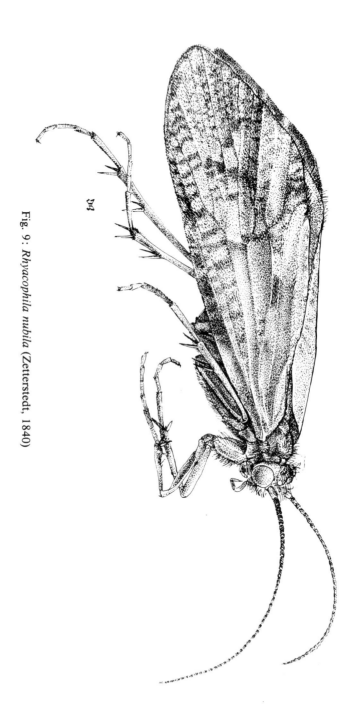

Fig. 9: *Rhyacophila nubila* (Zetterstedt, 1840)

majority restricted to cold, fast-flowing springs, streams and rivers of montane areas. The campodeiform larvae are large predatory animals, generally living on stony substrates, and not building cases; just before pupation, the last-instar larvae build large and strong pupal shelters from small stones, to protect the pupae in their leathery dark cocoons loosely attached to the inner walls of the stony shelters.

Key to the Species of Rhyacophila in the Levant

δδ

1. Dorso-apical lobe broad, rounded. Preanal appendages shorter than it, not deeply notched medially. Dorsal part of "body of segment X" protruding ventrad as a beak. Aedeagus and especially its ventral lobe much shorter than parameres, parameres not strongly elongated and with tendency to widening. Harpago high.   **R. fasciata** Hagen
2. Dorso-apical lobe narrow, tongue-shaped. Preanal appendages longer than it, deeply notched medially. Dorsal part of "body of segment X" not protruding ventrad as a beak. Aedeagus and its ventral lobe very long. Parameres strongly elongated, simply spiniform. Harpago distinctly longer than high.   **R. nubila** (Zetterstedt)

♀♀

1. A dark sclerotized trapezoid shield in the ventral split of the sclerotized collar of segment VIII.   **R. nubila** (Zetterstedt)
2. No sclerotized shield in the ventral split of the collar of segment VIII.

R. fasciata Hagen

Rhyacophila fasciata Hagen, 1859
Figs. 10–24

Rhyacophila fasciata Hagen, 1859, *Stettin. ent. Ztg*, 20:153. Schmid, 1970, *Mém. Soc. ent. Canada*, 66, Pl. VI, figs. 9–10.
Rhyacophila septentrionis McLachlan, 1865, *Trans. ent. Soc. London*, (3)5:157–158.
Rhyacophila aliena Martynov, 1915, *Izv. kavkaz. Muz.*, 9:2. Schmid, 1970, *Mém. Soc. ent. Canada.*, 66, Pl. VII:1–2. Nov. syn.
Rhyacophila talyshica Martynov, 1938, *Trudy zool. Inst. Baku*, 8:65–66, 72.

Wing expanse, in the very few winged specimens from Lebanon I could measure: δ 27.5 mm, ♀ 22.5–24.3 mm (these figures not very relevant). Wings: Fig. 18.
Male genitalia: segment IX with a dorso-apical lobe which is broad, more or less rounded (ovoidal). Plate formed by the two coalescent preanal appendages (beneath the dorso-apical lobe) shorter than this lobe, with apical margin not deeply notched medially in dorsal view. In lateral view, dorsal part of body of segment X (perpendicular on the above mentioned plate) strongly protruding ventrad as a beak. Aedeagus much shorter than parameres, with anteapical dorsal sharp point directed proximad, and with ventral lobe much shorter than itself. Parameres not strongly

21

elongated, sinuous, sometimes strongly broadened in the middle. Second segment (harpago) of gonopods large, rather high, with a more or less shallow apical emargination, separating a very faint superior lobe from a much more strongly protruding inferior lobe.

Female genitalia: At proximal part of segment VIII, a sclerotized collar, deeply and broadly split dorsally and ventrally; dorsal and ventral shape of this collar quite variable; but on the ventral side of the (less chitinized) more distal part of the segment VIII, always a pair of small elongate sclerites.

Remarks: *R. aliena* Martynov was determined by Dia & Botosaneanu (1983) from specimens caught in some small coastal rivers of Lebanon. Specimens sampled from the Orontes basin, also in Lebanon (Moubayed & Botosaneanu, 1985), showed important differences from the former specimens. A comparison of all the published documents on *R. aliena* and related taxa led to the conclusion that this is a synonym of *R. fasciata* Hagen. Since the limits of their variability clearly overlap in many cases, and because there seems to be no single character enabling the distinction of two taxa in all cases, there is no sound solution other than to synonymize *aliena* with *fasciata*. The variability of *fasciata* as well as that of *aliena* was already pointed out in several publications, and it certainly merits a thorough study, which will possibly lead to the distinction of several subspecies within the vast distribution area of this species (superspecies); for some of these, names are already available. This study is, of course, beyond the scope of the present publication. It should be noted that there are some authors who reject even the idea of the *fasciata-septentrionis* synonymy. Here are a few notes on the variability as shown by the Lebanese populations. This variability affects pratically all the parts of male and female genitalia. Comparison between the male genitalia of a specimen from the Nabaa Mourched spring in the Nahr el Aouali basin (Figs. 10–13) and those of a specimen from the Chlifa stream in the Orontes basin (Figs. 14–17) will reveal differences in the shape of the dorso-apical lobe of segment IX, of the latero-posterior margin of the same segment, in the dorsal shape of the "anal sclerites" (blackened in the figures), in the depth of the sinus separating the two apical lobes of the harpago, and also in the shape of the parameres, which are either narrow (like in the most "typical *fasciata*" from Europe), or markedly broadened (like in the most "typical *aliena*" from the Caucasus or Iran); moreover, the parameres are either covered with minute spines, or glabrous. A conspicuous variability affects also the dorsal and ventral shape of the sclerotized proximal collar of segment VIII of the female (compare Figs. 19 and 21 with Figs. 22–23), and sometimes the dorsal shape (Fig. 24, specimen from a small coastal river) strongly resembles that found in European *fasciata* but also in *aliena* from Iran, with medio-apical angles strongly produced (they are very faintly produced in specimens from another small Lebanese coastal river!). Recently published figures (Sipahiler, 1986) of male and female genitalia of "*R. aliena*" and "*R. fasciata*", respectively, from Anatolia, possibly point to a variability even wider than that shown by Lebanese specimens.

22

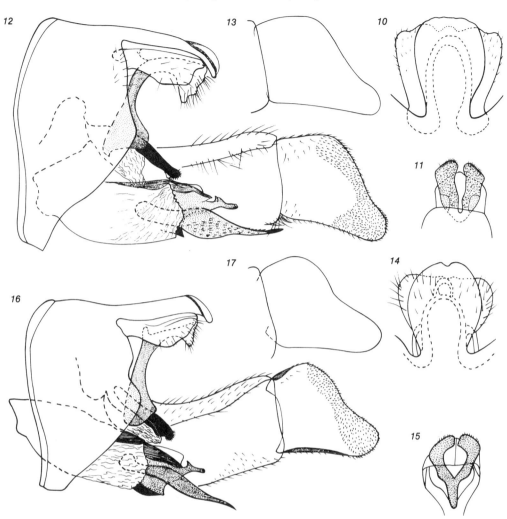

Figs. 10–17: *Rhyacophila fasciata* Hagen, 1859, male genitalia
10–13. specimen from Nabaa Mourched, Nahr el Aouali, Lebanon;
14–17. specimen from Chlifa, Orontes, Lebanon
(10 & 14. dorso-apical lobe of segment IX and preanal appendages, dorsal view;
11 & 15. "anal sclerites", dorsal view; 12 & 16. lateral view, left gonopod excised;
13 & 17. another lateral view of harpago)

Distribution: *R. fasciata* (in the present comprehensive sense, including *R. aliena*) has a very wide distribution in the western Palaearctic, being absent only from N. Africa, the Iberian Peninsula, and the northernmost parts; as *R. aliena*, it was initially described from the northern slopes of the Caucasus, and further determined from a vast territory including the Caucasus, Cis- and Transcaucasia, Asia Minor, Iran, and Lebanon. In the Levant it is known only from Lebanon, where it certainly reaches some of the southernmost points in its distribution (if not the southernmost ones),

23

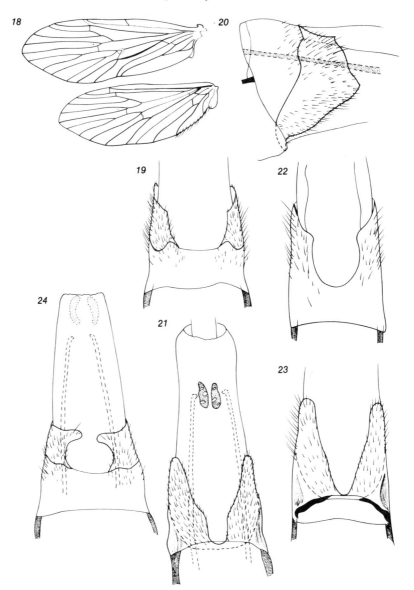

Figs. 18–24: *Rhyacophila fasciata* Hagen, 1859
18. wings;
19–21. female genitalia, and especially sclerotized proximal part of segment VIII
(19. dorsal view; 20. lateral view; 21. ventral view),
specimen from Nabaa Mourched, Nahr el Aouali, Lebanon;
22–23. sclerotized proximal part of segment VIII of female
(22. dorsal view; 23. ventral view),
specimen from Chlifa, Orontes, Lebanon;
24. segment VIII of female (dorsal view),
specimen from Nabaa el Barouk, Nahr el Aouali, Lebanon

its absence from Israel being virtually certain. The Lebanese localities presently known are either in the small coastal basins of Nahr ed Damour and Nahr el Aouali, or in the upper hydrographic basin of the Orontes.

Ecological Notes: In the mountain areas of Lebanon, the species inhabits springs, spring-brooks, and larger streams, but apparently never the main streams (and certainly never rivers) from ca. 700 m to ca. 1,400 m a.s.l. It is a psychrostenothermic species, restricted to colder, strongly flowing water. On the wing at least from April to December. It generally replaces *R. nubila* (Zetterstedt) in the colder, smaller watercourses of the Crenal and Rhithral, at higher altitudes, and coexistence of these two species must be an exceptional event.

Rhyacophila nubila (Zetterstedt, 1840)

Figs. 9, 25–28

Phryganea nubila Zetterstedt, 1840, *Ins. Lapp.*, p. 1068.
Rhyacophila nubila —. McLachlan, 1871, *Entomologist's mon. Mag.*, 7:281; Schmid, 1970, *Mém. Soc. ent. Canada*, 66, Pl. VIII, figs. 3–4.
Rhyacophila subnubila Martynov, 1934, *Opred. Faune SSSR* (*Tabl. analyt. Faune U.R.S.S.*), 13:42, 323.

Wing expanse (Levantine specimens): ♂ 19.5–25 mm, ♀ 22–26 mm. Habitus: Fig. 9.

Male genitalia: Segment IX with a dorso-apical lobe which is narrow, more or less tongue-shaped. Plate formed by the two coalescent preanal appendages (beneath the dorso-apical lobe) distinctly longer than this lobe, with apical margin deeply notched medially in dorsal view. In lateral view, dorsal part of body of segment X (perpendicular to the above-mentioned plate) not protruding ventrad as a beak. Aedeagus long, not much shorter than parameres, with a moderately developed anteapical dorsal point, and with long ventral lobe (not much shorter than aedeagus itself). Parameres simply spiniform, markedly elongated, covered with minute spines. Second segment (harpago) of gonopods small, distinctly longer than high, with moderately deep apical emargination separating a faint superior lobe from a much more strongly protruding inferior lobe.

Female genitalia: At the proximal part of segment VIII, a sclerotized collar, deeply split dorsally and ventrally. Medially placed in the ventral split is a sclerotized, dark, trapezoid shield.

Remarks: This species does not show conspicuous variability. Though this is not universally accepted, *R. subnubila* Martynov, 1934, described from the Caucasus, is almost certainly synonymous with *nubila*.

Distribution: *R. nubila* has a wide western Palaearctic distribution, but with some inportant gaps (absent from the British Isles, N. Africa, the Iberian Peninsula, the Pyrenees, the Alps; apparently very limited distribution in Italy); eastwards it is known from the Caucasus, from Asia Minor, and from northern Iran. In the Levant

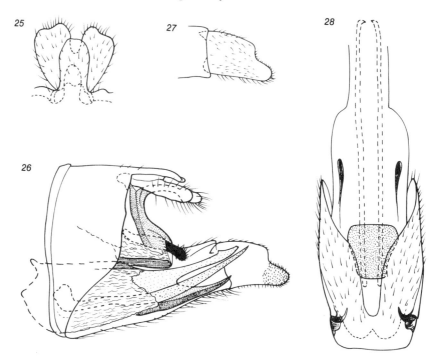

Figs. 25–28: *Rhyacophila nubila* (Zetterstedt, 1840)
25–27. male genitalia
(25. dorso-apical lobe of segment IX and preanal appendage, dorsal view;
26. lateral view, left gonopod excised; 27. another lateral view of harpago);
28. segment VIII of female, ventral view

it reaches (more exactly in Israel) the southernmost point of its distribution. It was caught in Lebanon in many localities in the hydrographic basins of two small coastal rivers — Nahr ed Damour and Nahr el Aouali — and also in the Orontes at Hermel, downstream of the Zarka spring. Without any doubt, present in South Lebanon, too. In Israel, the species occurs almost exclusively in the main watercourses forming the Jordan River at their junction: the rivers Dan, Senir (Ḥazbani), and Ḥermon (Banias) in Upper Galilee (1); it was also caught at Buteicha, near the north-east shore of Lake Kinneret (7), and at Senir, Golan Heights (18).

Ecological Notes: In the Levant the species essentially inhabits rivers and larger streams, from only a few tens of metres to ca. 800 m a.s.l. (it was rarely found also at ca. 950 m in the Lebanese mountains). It is a less psychrostenothermic and torrenticolous species than *R. fasciata* (in Israel it is restricted to fast flowing, cold streams, owing to ecological vicariance). Apparently on the wing throughout the year. It generally replaces *R. fasciata* in the warmer, larger watercourses of the Rhithral and Epipotamal, at lower altitudes, and coexistence of these two species, though reported, must be an exceptional event.

26

Family GLOSSOSOMATIDAE Wallengren, 1891

K. svenska VetenskAkad. Handl., 24(10):12, 163

Type Genus: *Glossosoma* Curtis, 1834.

This diagnosis is not valid for subfamily Protoptilinae, which is not represented in the Levant.

Small insects. Head with ocelli. Antennae relatively slender, shorter than forewings, scapus strong, but shorter than head. Maxillary palpi relatively setose, with 5 segments in both sexes, first two segments very short, last segment (like in labial palpi) less pointed than in Rhyacophilidae. Spurs: 2, 4, 4. Tibiae and tarsi of intermediate legs in female strongly dilated. Forewing elongated, rather regularly elliptical, covered by short, dense setae, with long fringes; venation complete or nearly complete: discoidal cell closed, forks 1–5 present. Hindwing narrower, discoidal cell closed or absent, forks 1, 2, 3, 5 or 2, 3, 5. Abdominal sternite V of male sometimes with very large lateral internal glands; sternites VI and VII with median "tooth". Male genitalia more simple than in Rhyacophilidae, sometimes asymmetrical, no intermediate appendages, gonopods unisegmented. Female with ovipositor like in Rhyacophilidae.

Key to the Genera of Glossosomatidae in the Levant

1. In hindwing f 1, 2, 3, 5, and discoidal cell closed; in ventral view ♂ genitalia asymmetrical. **Glossosoma** Curtis
– In hindwing f 2, 3, 5, and discoidal cell lacking; ♂ genitalia symmetrical. **Agapetus** Curtis

Genus AGAPETUS Curtis, 1834

Phil. Mag., 4:217

Type Species: *Agapetus fuscipes* Curtis, 1834; subsequent designation by Westwood, 1840.

Diagnosis: Small insects (Fig. 29), the smallest specimens being as small as most of the Hydroptilidae. Darkish, with very setose wings; setae not long, but both wings with long fringes and distinct venation. Antennae moderately stout, segments longer than broad, basal segment broader than the following and very short. Ocelli present, small, distinct. Maxillary palpus setose, stout, the two basal segments very short, 2nd segment globose, 3rd segment the longest, 5th segment shorter, 4th segment still shorter, but nevertheless clearly longer than 2nd. Spur formula (♂ , ♀): 2, 4, 4, spurs normal. Middle legs of female with very broad tibiae and tarsi, fringed with rigid setae. Forewings very setose, not very broad, regularly oval, with complete venation; hindwings shorter and distinctly narrower, venation somewhat reduced. In forewing discoidal cell closed and rather elongated, apical forks 1, 2, 3, 4, 5 present, apical cells

27

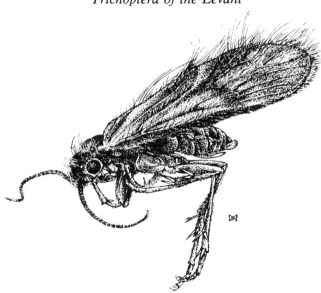

Fig. 29: *Agapetus caucasicus* Martynov, 1912, male

quite long (except forks 3 and 4, with long stalks); no closed median cell, long thyridial cell. Hindwings with very short subcosta rapidly joining the costal border, and not connected with radius by a transverse vein; the first following vein is, apparently, in its basal part the radius, and in its distal part the upper branch of sector radii (in some species, or specimens of some species, this ends in a short fork 1); no discoidal cell; forks 2, 3, 5 present, 2nd and 3rd with long stalks. At the level of abdominal sternite V in the male, two complex internal organs interpreted as glands, and a shorter or longer median appendage on sternite VI (in the female this appendage is present, but shorter than in the male). Male genitalia symmetrical, segment IX well sclerotized, preanal appendages present or absent, segment X forming a partly sclerotized and partly membranous roof above the phallic complex, one-segmented gonopods with various sets of spines or dentiform expansions. Female genitalia forming a short ovipositor, distal part of segment VIII a sclerotized ring, laterally compressed, segment IX apparently absent, segment X membranous but with a dorsal sclerotized structure, segment XI with a pair of bisegmented cerci (a different interpretation of the segments is possible).

General Remarks: *Agapetus* is a large genus with species in the Palaearctic, Nearctic, Oriental, and Australian regions (as well as in the high mountains of the Ethiopian region). Presently the genus is represented by two species in the Levant; the discovery of some additional species is plausible only for the northernmost parts of the province. The two known species belong to quite distinct species-groups. *Agapetus* are exclusively lotic insects, inhabiting especially springs and spring brooks, but many species are inhabitants of larger streams, a very fast current always being avoided. The eruciform larvae are typical grazers of algal vegetation on stony substrates, and construct extremely distinctive mineral cases which can be compared with tortoise shells (also having two dorsal openings). Before pupation, this complex construction is modified to a simple dome, its margins fastened to the substrate and sheltering the small, leathery pupal cocoon.

Key to the Species of Agapetus in the Levant

♂♂

1.   No preanal appendages. Segment X, in lateral view, broad also in its distal part, without bifid tips of its two halves. Apex of phallic complex deeply bilobed.
A. caucasicus Martynov

2.   Well developed preanal appendages. Segment X, in lateral view, strongly narrowing in its distal part, tips of its two halves clearly bifid. Apex of phallic complex not bilobed.
A. orontes Malicky

♀♀

1.   Sclerotized ring of segment VIII (lateral view) with distal margin simply oblique, without a sinus.
A. caucasicus Martynov

2.   Sclerotized ring of segment VIII (lateral view) with a deep sinus of distal margin.
A. orontes Malicky

Agapetus caucasicus Martynov, 1913
Figs. 29–34

Agapetus caucasicus Martynov, 1913, *Rab. Lab. zool. Kab. imp. varsh. Univ.*, pp. 20–22, Pl. II, figs. 5–6, Pl. III, fig. 6.

Wing expanse extremely variable, also within a single population; smallest ♂ measured: 5.1 mm, largest ♂ : 8.8 mm; smallest ♀ measured: 7.5 mm, largest ♀ : 9.6 mm; size larger in samples from Lebanon (♂♂ : 6.6–8.8 mm, ♀♀ : 7.8–9.6 mm) than in samples from Israel (♂♂ : 5.1–7.2 mm, ♀♀ : 7.5–8.7 mm). Habitus: Fig. 29. All parts of the body are brown, and some of them (head, meso- and metathorax) are dark brown, practically black. Forewing membrane moderately dark brown, the dense setal clothing is shiny, brown, giving the insect a blackish appearance. Wings: Fig. 30. In the male, on the level of abdominal sternite V, the usual pair of complex "glands" (also distinct in non-prepared specimens, as dark rounded areas); median appendage of sternite VI very long, extending beyond limit of segment VII. On abdominal sternite V of the female, a conspicuous semicircular thickened false suture; on sternite VI there is a moderately long median appendage.

Male genitalia: Seen from the side, segment IX with vertical proximal and distal margins. There are no preanal appendages. Segment X, from the side, very stout, apical part very obtuse, dorsal limit deeply emarginated; ventrally deeply split, with an important sclerotized reinforcement on each side, tips of these sclerites blunt.

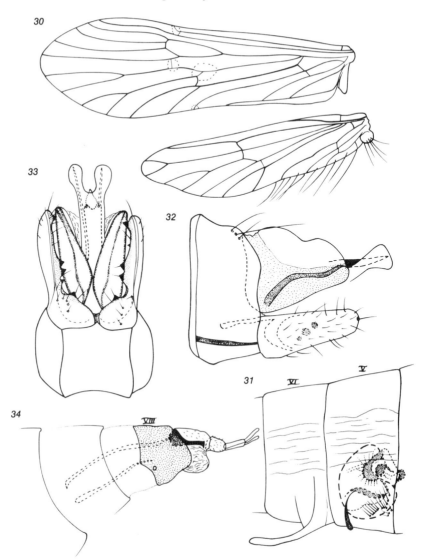

Figs. 30–34: *Agapetus caucasicus* Martynov, 1912
30. wings; 31. abdominal segment V and VI of male;
32–33. male genitalia (32. lateral view; 33. ventral view);
34. female genitalia (lateral view)

Beneath the roof formed by segment X is the phallic complex distinguished by its deeply bilobed apex (lobes large), and with a pair of slender parameres. One-segmented gonopods, very simple in lateral view, regularly and slightly narrowed towards a blunt apex; in ventral view they have a large median "heel" basally, from which the vertical, rather slender distal part of the gonopod arises; there is a variable armature of blackened teeth on the median edge of the heel (often only one), of the vertical part of the gonopod (mostly two larger teeth and some smaller), and near the apex (one).

Female genitalia: Seen from the side, the sclerite of segment VIII looks very simple, with a short projection of its anterior border, and with a simple, slightly oblique, posterior border; ventrally and apically, the segment is slightly split.

Remarks: *A. caucasicus* belongs to the *fuscipes* group of species (Ross, 1956), which includes about a dozen western Palaearctic species, almost all of them distributed around the Mediterranean. The variability of the species has already been mentioned.

Distribution: The species was described from Caucasus, and subsequently reported from northern Iran, Asia Minor, Cyprus, and Rhodes. It was mentioned also from "Mesopotamia", but this report, which lacks precision, cannot be trusted. It is plausible that it reaches the southernmost points of its distribution in the Galilee. In Lebanon it was caught in many localities in the hydrographic basins of small coastal rivers like Nahr ed Damour and Nahr el Aouali, fed by springs in the Barouk-Niha massif, and in the hydrographic basin of the Orontes. It will be certainly found everywhere in Lebanon and in the neighbouring parts of Syria, in suitable habitats. In Israel, all the known localities are in the extreme north of the country, in the Galilee (1) — including the foot of the Golan Heights; these are the springs, spring brooks and small streams at Tel Dan (Tel el Kadi), "Bet Ussishkin" (Nahal Layish), Nahal Hermon (Banias) and 'En Qedesh. Additional localities will no doubt be found here, but it is almost certain that the species does not extend further south.

Ecological Notes: In the Levant, *A. caucasicus* essentially inhabits springs of different types and spring brooks, from rather low altitudes to ca. 1,200 m a.s.l. at least. In suitable habitats of this kind it builds up enormous populations. It can also be found in somewhat larger streams, but in much smaller numbers, and is not present in the axial watercourses. The species is cold stenothermic and — though living only in running water — clearly avoids very strong current. On the wing apparently for the whole year (anyway, at least from January to October).

Agapetus orontes Malicky, 1980

Figs. 35–40

Agapetus orontes Malicky, 1980a, *Z. ArbGem. öst. Ent.*, 32(1/2): 1–2.

This species was caught in Syria ("Orontes bei Qssair"), thus belonging to the Levantine fauna. I can do nothing more than reproduce the text and illustration from the description by Malicky (1980a). In the original description, only the male was mentioned and described, but later (Malicky, 1983) the genitalia of a female considered as being that of *A.orontes* were also illustrated, without description, and I reproduce one of these drawings (Fig. 40).

"Körper, Anhänge und Flügel fahlgelblich, sklerotisierte Teile von Kopf, Thorax und Abdomen braun. Vorderflügellänge beim ♂ 4 mm. Kopulationsarmaturen ♂ : 9.

Segment überall ungefähr gleich breit, aber im Ventralteil nach kaudal zu verschoben. 10. Segment lang, in Lateralansicht mit gerader dorsaler und S-förmig geschwungener ventraler Kante. Distal endet das Segment jederseits in zwei kleinen Klauen, die nach unten und außen gebogen sind. In Dorsalansicht ist das 10. Segment bis über die Hälfte breit gespalten. Obere Anhänge lang und stumpf. Untere Anhänge lang und schlank mit ovalem Endteil. Von ventral gesehen verläuft ihre Innenkante bis knapp vor der Mitte gerade und parallel; von da an springt sie in einem rechten Winkel nach außen und verläuft dann im Bogen zur Spitze. Auf der Innenseite sitzen jeweils zwei kräftige sklerotisierte Kegel. Der Aedeagus ist lang und zylindrisch, dorsal mit einer sklerotisierten Längsleiste, im zweiten Drittel seiner Länge ist er am schmälsten.

Holotypus ♂ und 3 Paratypen ♂♂: Syrien, Orontes bei Qssair, 1.4.1979, leg. Kinzelbach, coll. Malicky.

Diese Art ist nächstverwandt mit *Agapetus laniger* PICTET, von diesem aber gut unterscheidbar: bei *laniger* sind 10. Segment und untere Anhänge noch viel länger, die unteren Anhänge sind nach unten gebogen, und die ventrale Innenkante der unteren Anhänge verläuft über ihre ganze Länge halbkreisförmig geschwungen ohne Vorsprünge und vorstehende Unebenheiten".

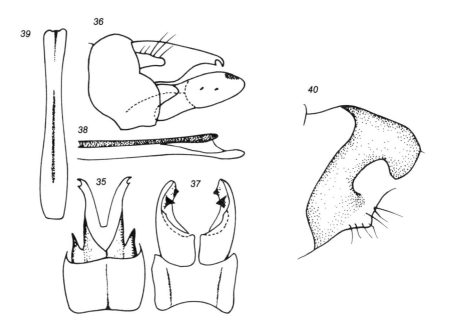

Figs. 35–40: *Agapetus orontes* Malicky, 1980
35–39. male genitalia
(35. dorsal view; 36. lateral view; 37. ventral view;
38. phallic apparatus, lateral view; 39. phallic apparatus, ventral view);
40. sclerotized ring of segment VIII of female, lateral view.
(35–39: from Malicky, 1980a; 40: from Malicky, 1983)

Genus GLOSSOSOMA Curtis, 1834

Phil. Mag., 4:216

Type Species: *Glossosoma boltoni* Curtis, 1834 (by monotypy).

Diagnosis: Medium-sized insects, generally light-brown, wings shiny, not very densely setose, with moderately long fringes, and with clearly distinct venation. Antennae shorter than wings. Ocelli present, large. Maxillary palpus: 1st segment short, 2nd even shorter and somewhat globose, 3rd longer than 1st + 2nd, 4th slightly longer than 1st and shorter than 5th (which ends in a minute point). Spur formula (♂, ♀): 2, 4, 4. Intermediate tibiae and tarsi in the female strongly dilated. Wings more or less regularly oval, but — especially forewing — with produced apex; the two pairs almost equally broad, but hindwings visibly shorter; venation complete, in both wings discoidal cell closed (rhomboidal or irregular, distinctly pointed at both ends), median cell open, thyridial cell long, narrow. Forewings: radius sharply bent before its end, at that point connected with the discoidal cell by a transverse vein, and then not very distinctly forked (lower branch thickened, pterostigmatic space betwen the branches of the fork darkened); forks 1, 2, 3, 4, 5 present; at the base of the male wing, a calosity which is sometimes covered with androconial scales or setae. Hindwings with forks 1, 2, 3, 5; radius parallel to subcosta, with distal part thickened. In the male, a large, flattened, median plate arising from sternite VI, and a broad triangular tooth from sternite VII, in the female only a broad tooth on sternite VI. In the different subgenera of *Glossosoma* the male genitalia are very diversely shaped; as both known Levantine species belong to sg. *Glossosoma* s.str., a general description could be as follows: genitalia asymmetric; sternite IX with an asymmetrical median appendage; segment X membranous, bordered by two diversely shaped sclerotized structures (which are sometimes called preanal appendages, and sometimes "sclerotized lobes of tergite X"); no intermediate, and no inferior appendages; complex phallic structures including: a basal "cup like" part, a pair of asymmetrical flaps arising from it, the phallus proper, and a very extensible " ventral process of the aedeagus" ending in a brush. In the female there is a long, slender ovipositor (segments VIII–XI), the last almost unmodified segment being segment VIII.

General Remarks: *Glossosoma* is not a very large genus, yet one displaying a remarkable diversity of features, especially of the male genitalia (the genus is split into a rather large number of subgenera; see, e.g., Ross, 1956). The species are distributed in the Palaearctic, Nearctic, and Oriental regions. In the Levant, two species are presently known, both belonging to sg. *Glossosoma* (with Palaearctic and Oriental representatives), and the discovery of some additional species in the area is not highly probable. *Glossosoma* are exclusively lotic insects, inhabiting various types of watercourses, but especially larger streams with a fast current and stony substrate. The feeding and building habits of the larvae and pupae are exactly like those of *Agapetus*; of course, larval and pupal constructions are distinctly larger.

Key to the Species of *Glossosoma* in the Levant

♂♂

1. Calosity at base of forewing without distal, freely moving ear-like flap. Process of sternite IX rather short, distinctly broader in its basal 3/4 than in its distal 1/4. "Preanal appendages" showing in their distal part a digitiform ventral process, widely separated from a short dorsal process. **G. hazbanicum** Botosaneanu & Gasith

2. Calosity at base of forewing with distal, freely moving ear-like flap. Process of sternite IX very long, uniformly slender. "Preanal appendages" showing in their distal part a ventral and a dorsal process which are both clearly hooked and not widely separated. **G. capitatum** Martynov

♀♀

1. Segment VIII (lateral view) proximally with an irregular, paler "sinus" (not always distinct!); bursa copulatrix with relatively complex internal, dark thickenings. **G. hazbanicum** Botosaneanu & Gasith

2. Segment VIII (lateral view) proximally paler and indistinctly sclerotized, but without "sinus"; bursa copulatrix with very simple internal, dark thickening. **G. capitatum** Martynov

Glossosoma hazbanicum Botosaneanu & Gasith, 1971

Figs. 41–50

Glossosoma hazbanica Botosaneanu & Gasith, 1971, *Israel J. Zool.*, 20:94–96. Type Locality: Hazbani, Israel. Type deposited in the Zoological Museum, Tel Aviv University.

Wing expanse: ♂ 9.2–15.8 mm, ♀ 14.5–17.4 mm; in the limits of a single population the size is not very variable, but there are important differences between different populations (e.g., in the ♂♂ examined from the Orontes at Hermel, the wing expanse varies between 9.2 and 10.2 mm; in those from Orontes-Yammouné, from 13.8 to 15.8 mm; in those from Nahal Senir [Hazbani] the range is 12–13 mm). All parts of the body are brownish, head and thorax are amber-like; antennae distinctly annulate with dark only in their proximal part; spurs darker than legs. Membrane of forewing shiny, iridescent, light brown, furnished with not very dense, short, generally golden setae, leaving the venation perfectly distinct. Hindwing resembling forewing in all these respects, somewhat paler; fringes of both wings dense and rather long. Venation of wings: Fig. 41; there is, at the base of the forewing, in the space between the 2nd and 3rd anal veins, a large horizontal (and flat) calosity, on which no androconial scales are inserted.

Male genitalia: Segment IX with very short sternite, and with a much better developed tergite, which is regularly produced (rounded) in the middle of its distal part; the appendage of sternite IX is not extremely long, shows a slight deviation to the right side, and is moderately narrow in its proximal 3/4, but suddenly narrows in its distal quarter, with the tip directed dextrally. Median part of segment X membranous, bilobed; it is bordered by the "sclerotized lobes of tergite X" (or "preanal

34

appendages"), whose apical part in lateral view shows the following elements: a "ventral process", long and slender (digitiform), curved ventrad and mediad; a "dorsal process" being a short tubercle (but pointed in dorso-ventral view); ventral and dorsal processes are largely separated by an oblique limit; a rounded extension follows immediately after the dorsal process, on the dorsal side. There are no inferior appendages (gonopods). The phallic apparatus is complex. Directly connected to the basal "cup-like" part of this complex is a pair of long pieces (quite distinct especially in ventral view) whose setose distal ends are asymmetrical: right one more slender than left one and also differing from it in shape. The apical part of the phallus proper is spoon-like (hollowed on the dorsal side), and slightly slit medio-apically; attached to the root of the phallus is the "ventral process of the aedeagus" (extremely extensible and ending in a strong, bifid brush of long black spines).

Female genitalia: In lateral view, the sclerotized ring of segment VIII shows an irregular sinus in the middle part of its proximal border; sternite VIII is less sclerotized (paler) along a narrow median part, and also at both ends. Bursa copulatrix (conspicuous, sclerotized internal structure of the ovipositor) curiously asymmetrical (twisted), with complex internal (median) thickenings.

Remarks: The size variability has already been mentioned; there is apparently no interesting variability of the male genitalia. The affinities of this species are not very clear, but it is possibly related to *G. discophorum* Klapálek and to *G. bifidum* McLachlan, the former being a Carpatho-Balkan species, the latter being known from the Alps, parts of the Balkans, and from the Mittelgebirge of Europe. *G. hazbanicum* is also related to *G. capitatum* Martynov (see below).

Distribution: Presently, *G. hazbanicum* is known only from Israel and Lebanon, being one of the possible endemic Trichoptera of the Levant. It was found in a number of localities in the hydrographic basins of the two small Lebanese coastal rivers, Nahr ed Damour and Nahr el Aouali, at ca. 1,000 m a.s.l. Moreover, it is present in the Bekaa Valley: upper course of the Orontes (where it was caught between 650 and 1,300–1,400 m a.s.l.), and also upper course of Yahfoufa, an affluent of the Litani, at 1,200 m a.s.l. It will certainly be found in many more suitable Lebanese localities, and probably also in the limitroph parts of Syria. In Israel, the species is exclusively known from the three large watercourses forming the Jordan at their confluence: Senir (Hazbani), Dan, and Hermon (Banias), at ca. 200 m a.s.l. (1)

Ecological Notes: The ecological spectrum of the species is certainly wider in Lebanon than in the Galilee. In the Lebanese mountains (at least from 650 to 1,400 m a.s.l.), medium-sized and larger streams with stony-sandy substrates are inhabited (many of them possibly belonging to the Metarhithral, but some of them to the Epirhithral), as well as large streams and their outlets; in suitable conditions, enormous populations are present. Several localities are known where the coexistence of *Glossosoma* with *Agapetus* has been established. In Israel (Galilee), where the species certainly reaches the southern limits of its distribution, it is restricted to larger, fast-flowing watercourses belonging to the Hyporhithral, and coexistence with *Agapetus* is no longer possible. *G. hazbanicum* was found on the wing throughout the year.

35

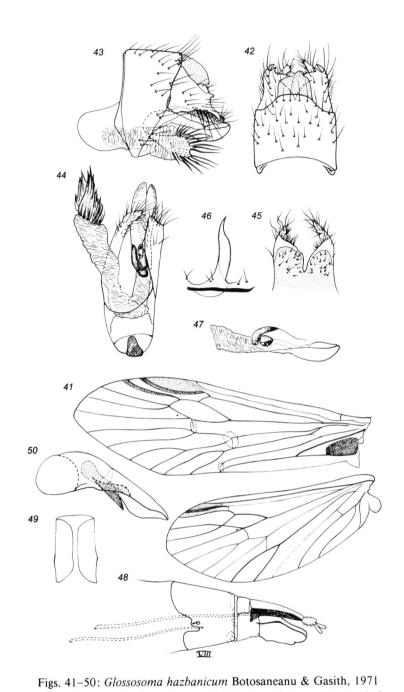

Figs. 41–50: *Glossosoma hazbanicum* Botosaneanu & Gasith, 1971
41. wings of male; 42–47. male genitalia (42. dorsal view; 43. lateral view;
44. phallic apparatus and its asymmetric lateral pieces, ventral view;
45. asymmetric lateral pieces of phallic apparatus and their common
membranous base, dorsal view; 46. appendage of sternite IX;
47. phallic apparatus, lateral view, more complete than in Fig. 43);
48–50. female genitalia (48. lateral view; 49. segment VIII, ventral view;
50. bursa copulatrix, lateral view). (42–47: from Botosaneanu & Gasith, 1971)

Glossosoma capitatum Martynov, 1913

Figs. 51–56

Glossosoma capitatum Martynov, 1913, *Rab. Lab. zool.Kab. imp. varsh. Univ.*, pp. 10–11, Pl. I, figs. 7–9.

This species was mentioned — without any details — in some recent publications (Kumanski & Malicky, 1984; Malicky & Sipahiler, 1984) as being found in Lebanon. This being plausible, it is here included. As no Levantine specimens were available, the drawings and description were prepared after 1 ♂ and 1 ♀ caught (7.VIII.1981) by Dr K. Kumanski (Sofia) in the Strandzha Mountains — oriental region of Bulgaria, and kindly made available to the author.

Wing expanse: ♂ 13 mm, ♀ 15 mm. Antennae distinctly annulate with brown in their basal 1/2. Thorax darkish brown. Membrane of fore- and hindwings of similar, pale brown color. At the base of the forewing, the space between 2nd and 3rd anal veins is highly modified by the presence of a calosity (without androconial scales, at least in our specimen preserved in alcohol) whose distal, freely moving part is conspicuously swollen, like a large, dark, vertical flap.

Male genitalia: Segment IX with very short sternite and with much better developed tergite, which is sharply produced distally, the distal border being somewhat irregular and extremely setose; the appendage of sternite IX is extremely long, it shows a distinct dextral deviation, and is very slender throughout its length. Median part of segment X membranous, bilobed; it is bordered by the "sclerotized lobes of tergite X"; at the apical part of these lobes, the "ventral process" as well as the "distal process" is distinctly pointed and hooked, in lateral as well as in dorsal view; the "ventral process" is much stronger than the "dorsal process", these two processes are not separated in lateral view by an oblique limit; a strong, rounded, very setose projection follows on the dorsal (and median) side after the dorsal process. No inferior appendages. Phallic apparatus apparently without interesting features (see *G. hazbanicum*). The long pieces connected to the basal "cup-like" part of the phallic complex have the same length and are almost symmetrical in the shape of their distal parts; nevertheless, the right one is straight, whereas the left one is definitely bent, its apical part forming an obtuse angle with the rest.

Female genitalia: In lateral view, the ring of segment VIII, although proximally paler, does not show a distinct sinus in the middle part of its proximal border; along the median line of sternite VIII, only a very narrow and incomplete, paler strip. Bursa copulatrix (conspicuous sclerotized internal structure of the ovipositor) symmetrical and rather slender, with very simple internal (median) thickening.

Remarks: Although clearly distinct from *G. hazbanicum*, this species is fairly close to it.

Distribution: *G. capitatum* is known from the Caucasus, Transcaucasia, Asia Minor and Lebanon. In the last mentioned country it is very probably restricted to the northern parts; here it probably reaches the southernmost limits of its distribution.

37

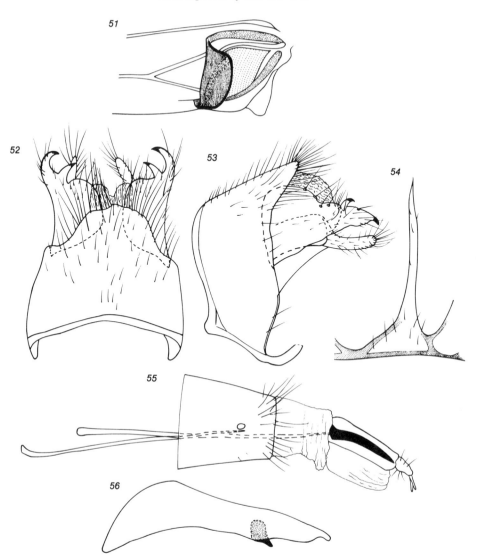

Figs. 51–56: *Glossosoma capitatum* Martynov, 1913,
from the Strandzha Mountains, Bulgaria
51. callosity at base of male forewing;
52–54. male genitalia
(52. dorsal view; 53. lateral view, without phallic apparatus; 54. appendage of sternite IX);
55–56. female genitalia
(55. lateral view; 56. bursa copulatrix, lateral view)

Family HYDROPTILIDAE Stephens, 1836
Ill. Br. Ent., 6:148, 151

Type Genus: *Hydroptila* Dalman, 1819.

Very small insects ("microtrichoptera"); but there are other caddisflies as small as, or smaller than larger Hydroptilidae. Distinctive characters are: the very dense covering of setae on all parts of body and especially on wings (completely obscuring venation), many setae on forewings being semi-erect and thickened; antennae always shorter or much shorter than forewings (shorter in female), with moniliform segments relatively constant in number in each species; scapus shorter than head. Ocelli present or absent. Maxillary palpi with 5 segments in both sexes, the first two very short, the last segment normal. Spurs long, formula varying in different genera, no more than one spur on foreleg. Wings generally narrow, even very narrow, apically pointed, with long or very long fringes (especially hindwings), venation mostly more or less markedly simplified, difficult to interpret. Abdominal segment V with minute lateral processes with apical setae or filaments. Sternite VII with median point or longer appendage. Male genitalia generally complex or very complex, sometimes with asymmetries or completely asymmetric, devoid of superior appendages. Female genitalia simple, shorter or longer ovipositor.

Key to the Genera of Hydroptilidae in the Levant

1. Large hydroptilids (wing expanse around 10 mm). Wings relatively broad, apex parabolic, venation fairly complete (f1, 2, 3 in forewing, and 1, 2, 3, 5 or 2, 3, 5 in hindwing). Ocelli present. Spurs 0, 3, 4.   **Agraylea** Curtis
   - Smaller. Wings narrower, apex more or less pointed, venation reduced (always less than 3 forks in either wing, f1 in hindwing always absent)   2
2. Spurs 0, 3, 4   3
   - Spurs 0, 2, 4 or 1, 2, 4   5
3. Ocelli present   4
   - Ocelli absent.   **Orthotrichia** Eaton
4. Wings very narrow, with long and sharp apices, venation indistinct and strongly reduced; sternite VII of ♂ and sternite VI of ♀ with median process.   **Oxyethira** Eaton
   - Wings less narrow, with less sharply produced apices, venation less simplified (forewing with f2, 3, hindwing with only f2). Sternite VI in both sexes with median process.   **Ithytrichia** Eaton
5. Spurs 1, 2, 4 (but spur of foreleg of ♂ minute!). Ocelli present. Completely black species, sometimes with silvery points.   **Stactobia** McLachlan
   - Spurs 0, 2, 4. Ocelli absent. Colour different.   **Hydroptila** Dalman

39

Genus STACTOBIA McLachlan, 1880

A Monographic Revision & Synopsis, pp. 505, 515–517

Type Species: *Hydroptila fuscicornis* Schneider, 1845; mentioned by Mosely (1933) as generotype; considered as type species by subsequent authors; selected as such by Fischer (1961).

Diagnosis: Very small insects, unmistakably recognized owing to the extremely setose, velvety-black wings, in many species sprinkled with silvery points which are distinct to the naked eye despite the minute size, as well as to their very special habitat and habits (see below). Ocelli present. Antennae always with 18 segments in both sexes — this constancy being exceptional even for Hydroptilidae, where in many species there is more or less constancy in the number of antennal segments. Maxillary palpi with first two segments minute, the following three of increasing length. Spur formula (♂, ♀): 1, 2, 4 (but the spur of the foreleg in the male is sometimes really minute, and said to be absent in one or two species). Both wings narrow, especially the hindwings; apices in most species distinctly pointed. The venation, rather uniform throughout the genus and with some distinctive features, will probably never supply good characters for distinguishing the species; it will suffice here to mention that forks 1, 2, 3 are present in the forewing, and forks 2, 3 (or only 3) in the hindwing. Abdominal segments densely clothed with black setae (the setae of the last abdominal segments have to be carefully removed, even in specimens macerated in KOH, in order to allow correct observation of the genitalia). On abdominal sternite VII, in the male, always a long, distally broadened appendage. Male genitalia simple, symmetrical (except for the phallus). Segment VIII somewhat modified, but with tergite and sternite always distinct, tergite smaller than sternite. Segment IX is represented virtually only by its tergite, which is always well developed, strongly sclerotized, forming a semi-cylindrical "roof", open ventrally; it has a pair of anterior extensions of its ventral-anterior angles, which can be very long; its ventro-apical angles are often produced in "apophyses". In direct relation with this tergite IX are three distinct formations; the most dorsal one is segment X with a central membranous part, bordered by sclerotized thickenings; beneath this are the sclerotized "superior appendages" (absent in a few species); and the lowest pieces, connected with the ventro-apical angles of tergite IX, are the movable inferior appendages, of very variable conformation. The phallus, with a dorsal position, very often has sclerotized internal spines of different shapes. In the female, sternite VI sometimes bears a small median tooth. The last unmodified segment is segment VII, all the following segments forming an oviscapt which is normally completely invaginated inside the preceeding segments (from which it has to be extracted with fine forceps in order to be observed); two pairs of very long and slender chitinous rods (apodemes) in this oviscapt may facilitate orientation: the dorsal pair belongs to segment X, the ventral one to segment VIII (sternite); a complex "vaginal apparatus" (normally inside the ovipositor, but very extensible) shows a sclerotized organ (bursa

40

copulatrix) which — in spite of the difficulty involved in observing it — probably offers the only distinctive specific characters for this sex.

General Remarks: There are presently slightly more than 80 known species of this remarkable genus; most of them belong to 6 species groups, but there are also some isolated species (Schmid, 1983). The largest number of species are Oriental; many others are western Palaearctic, mostly Mediterranean (*sensu lato*); and the exceptions are quite a few species on the Atlantic islands, in Africa, or in Arabia. Four species are presently known from the Levant, belonging to three distinct groups; it is almost certain that not only new localities for these species will be discovered here, but also some additional species, when their typical habitats will be systematically explored.

Many *Stactobia* (and possibly all the known Levantine species) are inhabitants of very distinctive habitats known as "hygropetric" or "madicolous": rocky surfaces permanently moistened by a thin layer of slow flowing water; on — or in the vicinity of these niches, the minute velvety-black insects run about swiftly in the sunlight, or fly awkwardly for very short distances. The larvae and pupae are to be found on the same rocky surfaces: the last instar larvae, with dark, heavily sclerotized plates on the thoracic and abdominal segments, build barrel-shaped cases (of secretion, sand, or calcite crystals), open at both ends.

Key to the Species of Stactobia in the Levant

♂♂

1. Phallus without any internal sclerotized formation; no superior appendages.
   S. pacatoria Dia & Botosaneanu
- Phallus with internal sclerotized "spines"; superior appendages present   2
2. Anterior extensions of segment IX very short; phallus with a pair of short, symmetrical black spines; superior and inferior appendages on each side forming a distinct forceps.
   S. margalitana Botosaneanu
- Anterior extensions of segment IX long and slender; phallus without a pair of symmetrical spines.   3
3. Anterior extensions of segment IX as long as the segment itself; phallus with only one, moderately stout spine; inferior appendages distinctly capitate and (laterally viewed) with a shaft.   **S. caspersi** Ulmer
- Anterior extensions of segment IX twice as long as the segment itself; phallus with two completely asymmetrical spines, one very long, one short, both twisted at base; inferior appendages not capitate, without shaft.   **S. aoualina** Botosaneanu & Dia

♀♀
(the ♀ of *S. margalitana* is presently unknown)

1. Bursa copulatrix bifid at distal end.   **S. caspersi** Ulmer
- Bursa copulatrix not bifid at distal end   2
2. Distal end of bursa copulatrix not sclerotized, pale.   **S. aoualina** Botosaneanu & Dia
- Distal end of bursa copulatrix heavily sclerotized (dark), capitate and hollow.
   S. pacatoria Dia & Botosaneanu

41

Stactobia caspersi Ulmer, 1950
Figs. 57–64

Stactobia caspersi Ulmer,1950, *Arch. Hydrobiol.*, 44:294–300. Schmid, 1983, *Naturaliste can.*, 110:282
Stactobia eretziana Botosaneanu & Gasith, 1971, *Israel J. Zool.*, 20:96–98. Type Locality: 'En 'Avedat, Israel. Type deposited in the Zoological Museum, Tel Aviv University.

Wing expanse: ♂ 4.2–5.1 mm, ♀ 5.3–5.5 mm (one ♀ with a wing expanse of 4.6 mm was exceptionally found). Wings black (the specimens being preserved in alcohol, it is not known if there are silvery points on the forewings). Wings: Fig. 57.

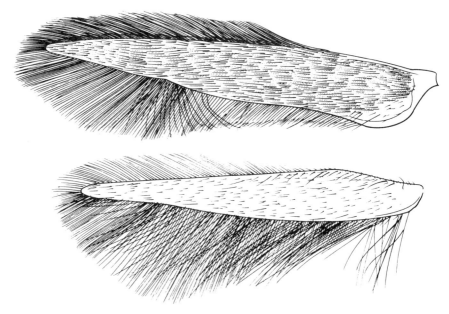

Fig. 57: Wings of *Stactobia caspersi* Ulmer, 1950 (venation not represented)

Male genitalia: Ventral appendage of segment VII extending slightly beyond middle of sternite VIII. Moderately large segment (or tergite) IX, apically broadly rounded, without "apophyses" of its ventro-apical angle; the slender anterior processes are as long as the segment proper. Segment X with poorly developed, bilobed membranous part, but with well developed "dorsal thickenings" (in dorsal view: two parallelogrammic plates) which are laterally bordered with black (this is easily seen in lateral view also). Laterally, beneath segment X, are the superior appendages: rather small, obtuse at end, with a hollow on their inferior side — into which the superior angle of the inferior appendages penetrates. The inferior appendages (gonopods) are perhaps the most conspicuous part of the genitalia: they are strongly capitate and with a moderately long and strong "shaft"; in ventral view the "head"

42

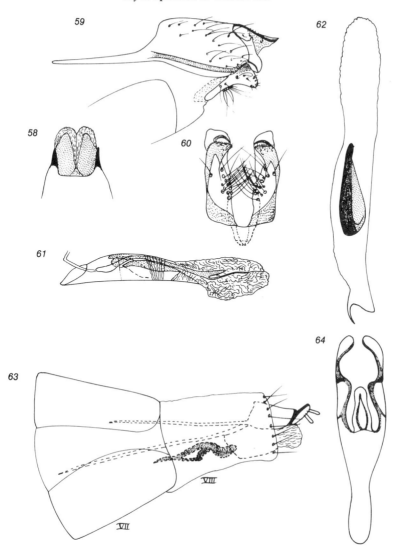

Figs. 58–64: *Stactobia caspersi* Ulmer, 1950
58–62. male genitalia
(58. segment X, dorsal view; 59. lateral view;
60. inferior appendages, posterior view;
61. phallic apparatus, lateral view; 62. sketch of phallic apparatus, ventral view,
showing the heavily sclerotized internal spine);
63–64. female genitalia
(63. lateral view; 64. bursa copulatrix, ventral view).
(58–61: from Botosaneanu & Gasith, 1971)

is clearly projecting mediad, and there is an important zone furnished with long setae beneath this "head", on the median face of the gonopod. Phallus has inside (besides a long and slender, mostly sclerotized dorsal structure connected with the ductus ejaculatorius, and which should not be confused with a spine), a unique heavily sclerotized spine which is rather short and broad, and only very slightly curved.

Female ovipositor represented in Fig. 63 (internal sclerotized structure of the "vaginal apparatus": Fig. 64).

Remarks: *S. caspersi* belongs to a large species-group (group *furcata*), presently including some 20 species (Schmid, 1983), many of them Mediterranean, but some of them occurring on the Atlantic islands, or penetrating into Central Europe, one being present in north Iran. The species is not very variable, despite its rather wide distribution; nevertheless, it may be noted that in the specimens from Bulgaria (Ulmer, 1950; Kumanski, 1979) the internal phallic spine is certainly longer and more curved than in Levantine specimens; there are also slight differences in the gonopod shape.

Distribution: *S. caspersi* was described from madicolous habitats in the vicinity of Varna (Bulgaria), and has subsequently been discovered in Italy, the mountains of Banat (Romania), Bulgaria, Greece, and from various eastern Mediterranean islands. In the Levant, small numbers were sampled in two Lebanese localities on small coastal rivers (Nahr el Hammam, a tributary of Nahr ed Damour, at 45 m a.s.l. and 7 km from the sea; Nahr el Aouali at 230 m a.s.l. and about 30 km downstream from the head of this river). In Israel, a huge population was observed and sampled at a large madicolous habitat with slightly brackish water, alongside the desert stream of 'En 'Avedat, at 475 m a.s.l. in the Central Negev (17). It is certain that the species will be found in other madicolous habitats throughout the Levant, being here supposedly the most frequent representative of the genus.

Ecological Notes: Typical madicolous species. Found on the wing at least from January to September.

Stactobia margalitana Botosaneanu, 1974

Figs. 65–66

Stactobia margalitana Botosaneanu, 1974b, *Israel J. Ent.*, 9:168–170. Type Locality: Nahal 'Arugot, in the Dead Sea Depression, Israel. Type deposited in the Zoological Museum, Tel Aviv University.
Stactobia margalita (sic!) —. Schmid, 1983, *Naturaliste can.*, 110:282.

Wing expanse, ♂ : 3.2 mm (an extremely small species). As specimens are preserved in alcohol, it is not possible to know if there are silvery points on the forewings.

Male genitalia: There is no ventral appendage on segment VII. Segment (or tergite) IX with very short anterior processes; the tergite becomes curiously narrower (=lower) posteriad; there is a well-developed "apophyse" of its ventro-apical angle (this apophysis is, in lateral view, bilobed, the inferior lobe blunt, the superior lobe pointed). Segment X with median membranous part bordered by a pair of sclerites (the "dorsal thickenings") which are symmetrical and of characteristic shape (see lateral figure). Superior and inferior appendage on each side forming an easily discernible forceps. Superior appendages large, not strongly sclerotized, in lateral view simple, more complex in ventral view, where their rounded apices come into contact on the median line. Inferior appendages bilobed; this is less distinct in lateral view (where only the long inferior lobe is easily seen, with its narrowed tip directed posteriad), but very distinct in ventral view. Phallus with a pair of symmetrical internal spines, very dark, short, slightly convergent apically.

Female unknown.

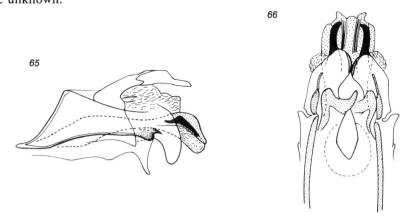

Figs. 65–66: *Stactobia margalitana* Botosaneanu, 1974, male genitalia
65. lateral view; 66. ventral view (median lobe of right gonopod not represented).
(From Botosaneanu, 1974b)

Remarks: *S. margalitana* belongs to the *furcata* species-group (see *S. caspersi*), being perhaps more closely related to *S. eatoniella* McLachlan, from the western European mountains.

Distribution and Ecological Notes: Presently, this species is known only from the "locus typicus": this is the valley of the desert stream Naḥal 'Arugot, in the Dead Sea Depression, near 'En Gedi, at ca. 350 m below sea level (13). Only two (male) specimens were caught there, on 23rd June 1970 and 26th April 1984, respectively. No specimens were caught at a madicolous habitat, but this probably means merely that the true habitat of the species in Naḥal 'Arugot has yet to be discovered.

Stactobia aoualina Botosaneanu & Dia, 1983

Figs. 67–71

Stactobia aoualina Botosaneanu & Dia, in: Dia & Botosaneanu, 1983, *Bull. zool. Mus. Amsterdam*, 9(14):127–128. Type Locality: Nabaa Aazibi, on the tributary Nahr Aaray of

Nahr el Aouali, Lebanon. Type deposited in the Zoological Museum of the University of Amsterdam.

Wing expanse: ♂ 4.3–5.1 mm, ♀ 4.5–4.7 mm. Wings black (the specimens being preserved in alcohol, it is not known if there are silvery points on the forewings). Male genitalia: Long ventral appendage of segment VII, extending to 3/4 of the length of sternite VIII. Tergite IX trapezoidal, inferior edge oblique, anterior processes twice as long as the tergite itself, ventro-apical angles not very produced, somewhat irregular; adhering to these angles are the superior appendages, having the appearance of swollen knobs, almost touching on the median line in ventral view. Inferior appendages moderately elongated, nearly parallel, tips obtuse with a pair of spines inserted nearby. Segment X not very characteristic. Phallus with structure correctly represented in Fig. 69, and with two completely asymmetric internal spines: distal spine short, extremely twisted, sharp tip directed basad; proximal spine extremely long, proximally twisted, blunt tip directed distad.

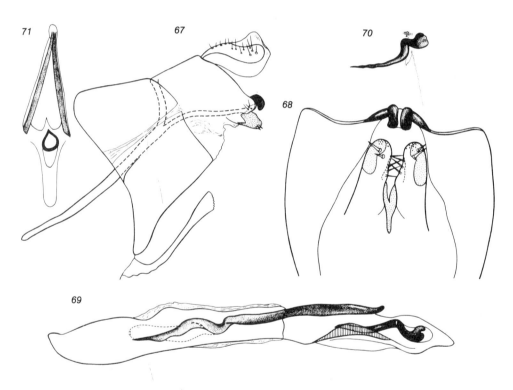

Figs. 67–71: *Stactobia aoualina* Botosaneanu & Dia, 1983
67–70. male genitalia
(67. lateral view; 68. superior and inferior appendages, ventral view;
69. phallic apparatus, lateral view; 70. distal spine of phallic apparatus, ventral view);
71. bursa copulatrix of female.
(From Dia & Botosaneanu, 1983, except 71)

46

Internal sclerotized structure ("bursa copulatrix") of the "vaginal" apparatus" of the female: Fig. 71.

Remarks: *S. aoualina* belongs to the *martynovi* species-group, including more than 20 known species, most of them Oriental, the few others inhabiting Cyprus, Anatolia, northern Iran or the Far East; *S. aoualina* seems to be related to a small complex of non-Oriental species, and also to a species from Tien Shan.

Distribution and Ecological Notes: There are two known localities for this species, in the hydrographic basin of the small Lebanese coastal river Nahr el Aouali and in the Niha massif: Nabaa Aazibi (springs feeding the stream Nahr Aaray, or Jezzine, at 900 m a.s.l.) and Nabaa Abou Kharma (spring and spring-brook between the villages Haret Jandal and Bâter ech Chouf, 850–760 m a.s.l.). No exact observations were made pointing to the hygropetric character of the habitats. The adults were caught in June, August and October.

Stactobia pacatoria Dia & Botosaneanu, 1980
Figs. 72–78

Stactobia pacatoria Dia & Botosaneanu, 1980, *Bijdr. Dierk.*, 50(2):369–374. Type Locality: madicolous habitat, stream Ouadi Ras el Mâ, near village Haret Jandal ech Chouf (Lebanon, hydrographic basin of Nahr el Aouali). Type deposited in the Zoological Museum of the University of Amsterdam.

Wing expanse: ♂ 4.6 mm, ♀ 4.8–5.2 mm. Wings black, no silvery points on the forewings. Spur of ♂ foreleg extremely small.

Male genitalia: Moderately long ventral appendage of segment VII, with lower margin non-serrate, extending beyond the middle of sternite VIII. Tergite VIII with posterior edge extremely oblique; sternite VIII, in ventral view, with deep triangular split of its apical edge. Segment (tergite) IX with anterior processes longer than itself; posterior edge very oblique, ventro-apical angles produced in moderately well developed obtuse protuberances. Segment X stout, strongly developed and quadrangular, upper edge horizontal, not strongly narrowed apically, with central dorsal zone feebly sclerotized and imperfectly limited, apically furnished with many small spines. There are no superior appendages. Inferior appendages conspicuous, elongated, heavily sclerotized, covered with small spurs, hanging ventrad between the segments IX and X; ventrally they are more or less triangular, each with a very strong "heel", not coalescent on the median line, but instead, almost completely separated by a very long and rather broad distal sinus, and by a strongly developed proximal "bay" (though united to a common "root" which is extremely long — as long as the inferior appendages themselves). Phallus a slender tube without internal spines; apex sinuous and pointed in lateral view, somewhat lanceolate in dorso-ventral view.

Internal sclerotized structure ("bursa copulatrix") of the "vaginal apparatus" of the female (Fig. 78), with tip heavily sclerotized, dark brown, capitate and hollow, easily visible in the abdomen macerated in KOH.

47

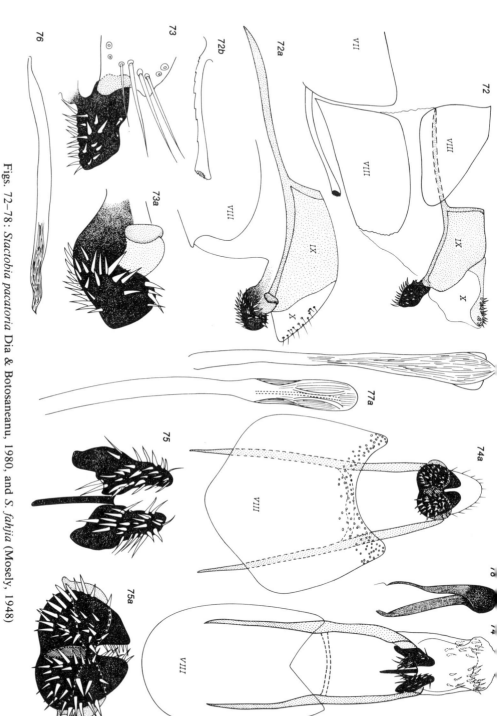

Figs. 72–78: *Stactobia pacatoria* Dia & Botosaneanu, 1980, and *S. fahjia* (Mosely, 1948)

72–77. male genitalia; numbers without letters: *S. pacatoria*; numbers followed by a letter: *S. fahjia*

(72, 72a & b, lateral view; 73, 73a, inferior appendage, lateral view, more strongly enlarged;

74, 74a.: ventral view; 75, 75a. inferior appendage, ventral view, more strongly enlarged;

76. phallic apparatus, lateral view; 77, 77a. phallic apparatus, ventral view); 78. bursa copulatrix of female *S. pacatoria*.

(72–77: from Dia & Botosaneanu, 1980)

48

Remarks: This species belongs to the *nielseni* species-group, presently including 11 species, most of them Oriental. *S. pacatoria* is closely related to a species known from Aden (South Yemen) — *S. fahjia* (Mosely, 1948:76–77) — but clearly distinct from it in many details. The presence of this pair of sister species separated by some 2,600 km is an interesting situation.

In fact, the differences between males of the two species are evident from a careful comparison of the two original illustrated descriptions. I have, nevertheless, studied several male paratypes of *S. fahjia* from Jebel Jihaf, Wadi Leje (kept in the BMNH; courtesy of Dr P.C. Barnard), and I have prepared Figs. 72a, b, 73a, 74a, 75a and 77a from them. There are differences in the shape of segment X (only moderately developed, triangular, and with finer setae, in *S. fahjia*; but it is not impossible that segment X is, to a certain extent, erectile). The very broad sternite VIII has, in this species, a broad and deep, perfectly rounded distal emargination; the lower margin of the ventral appendage of segment VII is distinctly serrate; the ventro-apical angles of tergite IX are produced in strong protuberances with a rather complex relief, leaning on the dorsal side of the gonopods, and distinct laterally as well as ventrally. But the most conspicuous differences are in the structure of the inferior appendages, which, in *S. fahjia*, are less elongated (more globose), to a large extent coalescent, separated only by a relatively short and narrow distal sinus (which may be more triangular than in Fig. 75a), whereas proximally there is only an extremely slight emargination; moreover, the common "root" of the gonopods is short, much shorter than the inferior appendages themselves, and not protruding proximally beyond them.

Distribution and Ecological Notes: The only known locality is a vast madicolous habitat in the bed of the Ouadi Ras el Mâ stream, near the village of Haret Jandal ech Chouf (Lebanon, catchment of the coastal river Nahr el Aouali, 800–900 m a.s.l.). The habitat is typically hygropetric with very hard water. Some adults were on the wing in mid March, but it could be inferred that the peak (or one of the peaks) of imaginal emergence is in April–May. The population is very large.

Genus AGRAYLEA Curtis, 1834[*]
Phil. Mag., 4:217

Type Species: *Agraylea sexmaculata* Curtis, 1834 (subsequent designation by Westwood, 1840). The type species of subgenus *Allotrichia* is *Agraylea pallicornis* Eaton, 1873 (by monotypy).

Diagnosis: Very large (the largest) hydroptilids, wing expanse ca. 10 mm (sometimes larger). This means that they are larger than some non-hydroptilids (in the Levantine fauna, e.g., *Agapetus, Ecnomus, Psychomyia, Lype*). Ocelli present. In the antennae,

[*] Including *Allotrichia* McLachlan, 1880. See General Remarks on the genus, below.

of the male, 27 is the most frequently encountered number of segments (sometimes one or two less in the female). Wings relatively broad, apices more or less parabolic, never very pointed — even in the hindwings; venation fairly complete (forks 1, 2, 3 in forewing, and 1, 2, 3, 5 or at least 2, 3, 5 in the hindwing). Spur formula: 0, 3, 4. Generally a long median process on abdominal sternite VII in the male, and a minute tooth on sternite VI of the female. Male genitalia very difficult to describe in general terms; segment IX usually reduced to a rudiment on the dorso-median line; segment X with a dorsal membranous part, and with a sclerotized ventral part, generally resembling a protruding lobe, with or without processes of various shapes; inferior appendages well developed but not very protruding, of varied shapes; phallic apparatus with distinct proximal and distal part, with a coiled paramere. Female genitalia forming an oviscapt, which is apparently quite differently formed in the two subgenera (but this has to be verified!). In sg. *Allotrichia* the last normally developed segment is segment VII; segment VIII is much narrower but well developed, with a conspicuous crown of long, black apical setae ventrally, and in some species with a longitudinal sclerotized keel on the sternite, and segments IX–XI can be well distinguished (although they are normally completely telescoped within segment VIII). In *Agraylea* (s. str.), according to Schmid (1980:38, Figs. 36–37), the oviscapt has a quite different structure, with a strongly modified segment VII, its sternite bearing a crown of apical setae; the existence of segment VIII (and IX?) is questionable.

General Remarks: It is evident that there is still much imprecision in our knowledge concerning *Agraylea* — either s. str. or s. lat. This is a very small genus, even in the broader sense adopted here, including less than a dozen known species, some Palaearctic, some Nearctic, one reputedly Holarctic. Quite a few of these species are widely distributed, but most of them have a sporadic and/or restricted distribution.

Marshall (1979:193–197) has well summarized the serious doubts concerning the validity of the genus *Allotrichia* McLachlan, 1880 as distinct from *Agraylea*, and such doubts were also expressed by other authors. To be concise: I agree that *Allotrichia* merits at most the status of a subgenus of *Agraylea*. The only imaginal non-genital character apparently always distinguishing *Agraylea* (s. str.) from *Agraylea* (sg. *Allotrichia*) is the presence in the forewing of *Agraylea* (s. str.) of fork 1 much longer than fork 2 (in other words, the upper branch of sector radii furcates much earlier than the lower branch), whereas in sg. *Allotrichia* the two branches of sector radii furcate at about the same level. In the literature, mention was often made of a second distinctive character in the venation, the absence of fork 1 in the hindwing of *Allotrichia* and its presence in *Agraylea*. However, this is an unstable character, several instances being known of *Allotrichia* with this fork. It is possible that some good male genital characters will be found to distinguish the two subgenera (I am moderately sceptical about this). As for the female oviscapt, it is really possible that good distinctive characters exist (see Diagnosis).

The last instar larvae build opaque, laterally compressed cases resembling spectacle-cases, filamentous algae always being an essential item used in the construction; the pupal case is the closed larval one, attached parallel or perpendicular to the substrate by 4 or 2 adhesive pedicels.

In the Levant, as yet only two representatives of sg. *Allotrichia* are known, but it is almost certain that a 3rd one belongs to its fauna (*pallicornis* Eaton, 1873; originally described as *Agraylea*(!); rather widely distributed in Europe, North Africa, Asia Minor, and north Iran; this species has been included in the keys). Moreover, it is not impossible that some *Agraylea* (s.str.) will also be discovered here at some time, this being more plausible for *A. (A.) sexmaculata* Curtis, 1834, which has a wide western Palaearctic distribution.

Key to the Species of Agraylea (sg. Allotrichia) in the Levant

♂♂

1. Distal border of segment IX quite regular, oblique; segment X (lateral view) pointed at apex, curved ventrad; inferior appendages irregularly quadrangular from side, tips obtusely rounded in dorso-ventral view, without important emargination, without "teeth" on median side. **A. teldanica** (Botosaneanu)

– Distal border of segment IX irregular; segment X (lateral view) obtuse at apex; inferior appendages distinctly emarginated (lateral or dorso-ventral view), with "teeth" on median side   2

2. Inferior appendages (from side, also dorso-ventrally) with distal excision defining a superior (outer) and an inferior (inner) lobe of about the same size; the symmetrical spiniform appendages of segment X, seen from side, are, generally speaking, directed upwards. **A. pallicornis** Eaton

– Inferior appendages (from side, also dorso-ventrally) with distal excision defining two lobes, superior (outer) lobe distinctly more produced than inferior (inner) lobe; the asymmetrical spiniform appendages of segment X, seen from side, are, generally speaking, directed posteriad. **A. vilnensis orientalis** ssp. nov.

♀♀

1. Sternite VIII without longitudinal keel, with a longitudinal groove. **A. pallicornis** Eaton
– Sternite VIII with a longitudinal sclerotized median keel   2
2. Keel of sternite VIII not very distinctly capitate, broad, with median longitudinal darker zone. **A. teldanica** (Botosaneanu)
– Keel of sternite VIII very distinctly capitate, slender except for the ends, "head" distinctly darker than the rest. **A. vilnensis orientalis** ssp. nov.

Agraylea (Allotrichia) teldanica (Botosaneanu, 1974)

Figs. 79–84

Allotrichia teldanica Botosaneanu, 1974b, *Israel J. Ent.*, IX:164–168. Type locality: Tel Dan (Tel el Kadi), Israel. Type deposited in the Zoological Museum, Tel Aviv University.

Wing expanse: ♂ 8–10.5 mm, ♀ 8.4–11 mm (the specimens from Lebanon are somewhat larger than those from Israel). Antennae with 27 segments in the ♂, with 25–26 in the ♀. Head and thorax brown to dark brown, legs and abdomen less dark. Wings: Fig. 79; both of them, but especially forewings, very densely clothed with woolly, dark brown setae, with scattered erect black setae, fringes very dense, not extremely long; this renders observation of the venation, which is anyway very indistinct, quite difficult; Fig. 79 will, nevertheless, give a good image of this venation, which is typical for the genus (with forks 1 and 2 in the forewing starting at about the same level, and with hindwings devoid of furca 1). On abdominal sternite VII of the male, a sinuous median appendage, almost reaching the middle of the next tergite; this appendage has a broadened, semicircular and concave apex. On abdominal sternite VI of the female, a minute median tooth.

Male genitalia: Segment IX with the tergite almost interrupted medially; the lateral shape of the segment is characteristic, with, for instance, a smooth, oblique posterior edge and with postero-ventral angles produced in prominent rounded lobes. Segment X with central zone of its proximal part membranous, followed distally by a large lobe bordered by the two strongly sclerotized "roots" of the long spiniform appendages characteristic of most of the species of this genus; the central lobe is long, reaching beyond the tips of the inferior appendages; seen laterally, it is curved ventrad, with apex clearly pointed; it is hollow ventrally, forming a "roof" for the phallus. The spiniform appendages are practically symmetrical, with bulbous basal part directed basad and laterad, the spine itself being directed first ventrad and posteriad, then clearly dorsad and mediad. Inferior appendages (gonopods): in dorso-ventral view very simple, divergent, with simply rounded apices; in lateral view irregularly rectangular, with slightly produced postero-ventral angle; there is no tendency towards a bilobed shape, and there are no teeth on the median side.

Female genitalia: Medially on sternite VIII is a longitudinal sclerotized keel, almost reaching the end of this sternite; it is broad, not very distinctly capitate, and characterized by a median longitudinal zone which is clearly darker than the lateral zones.

Remarks: *A. teldanica* is not closely related to any of the other known species of the genus. There is as yet no evidence of some intraspecific variability.

Distribution: There are mentions of this species from some Aegean islands and from Asia Minor (Botosaneanu & Malicky, 1978; later on, in Malicky & Sipahiler, 1984, it is no longer mentioned from Turkey). The known Levantine localities are in Lebanon and in Israel. In all of them quite small numbers of specimens were sampled, pointing to a sporadic distribution. In Lebanon only 3 specimens were caught, in two

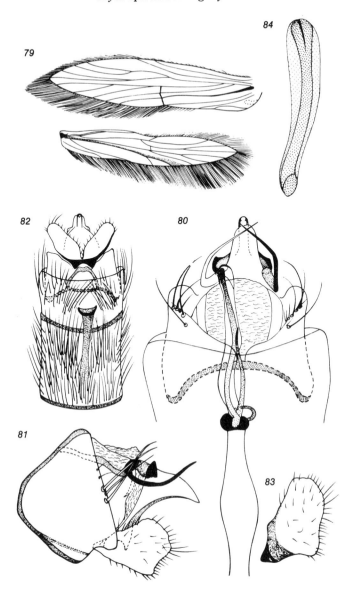

Figs. 79–84: *Agraylea* (*Allotrichia*) *teldanica* Botosaneanu, 1974
79. left forewing, right hindwing;
80–83. male genitalia
(80. dorsal view; 81. lateral view; 82. ventral view;
83. gonopod, lateral view, slightly different angle);
84. longitudinal keel on sternite VIII of female, ventral view.
(From Botosaneanu, 1974b, except 84)

localities in the hydrographic basin of the small coastal river Nahr el Aouali, in the Niha massif (Nabaa Bâter ech Chouf, at 820 m a.s.l. and Nabaa Aazibi, at 900 m a.s.l.). In Israel, this species was caught in the important complex of springs and streams of Tel Dan, in the Galilee, at 200 m a.s.l. (1). *A. teldanica* will certainly be caught at other Lebanese localities, throughout the country; it seems less probable that it will be discovered in Galilean localities other than Tel Dan.

Ecological Notes: The poor information available may be summarized as follows. In the Levant the species seems to be a crenobiont (or at least a crenophilous species), all the known localities being springs or spring brooks. Some of the specimens were caught in light traps. *A. teldanica* is on the wing at least from April to September.

Agraylea (Allotrichia) vilnensis orientalis ssp. nov.

Figs. 85–87

Allotrichia vilnensis Raciecka, 1937. Schmid, 1959, *Beitr. Ent.*, 9(5/6):685–686; Dia, 1983, Rech. écol. biogeogr. des cours d'eau du Liban méridional, pp. 133, 189. Type kept in the Zoological Museum, University of Amsterdam.

Preliminary Note: The following description is not really satisfactory; that of the female is based on the type from Lebanon; but that of the male is based on the description by Schmid (1959) of specimens from Iran. Nevertheless, 1 male from the same locality as the female type — and, unfortunately, not available for the present description — was earlier compared by me with the description by Schmid, and found to agree quite well with it. A description of *Allotrichia vilnensis* from Bulgaria was published too late to be used for the present book.

Wing expanse: ♀ 10.6 mm. Antennae with 27 segments (♀). Head and thorax dark brown. Forewing very densely clothed with woolly, dark brown setae, completely obscuring the venation, and with numerous erect, black setae, scattered over the wing; pterostigmal zone distinctly darker than the rest of the wing; hindwings somewhat paler. In the female, forks 1, 2, 3 in the forewing, and 2, 3, 5 in the hindwing (in the hindwing, the upper branch of SR is connected to R1 through an oblique transverse vein arising near the forking point of SR, as in *A. teldanica*). On abdominal sternite VII of the male, a long appendage; on sternite VI of the female, a minute tooth.

Male genitalia: Segment IX stout, in lateral view with very irregular distal limit, with two rather deep emarginations, postero-ventral angles only moderately produced. The lower complex of segment X represented by a conical central lobe which does not reach the tips of the inferior appendages, and by a pair of definitely asymmetrical lateral lobes, each of them bearing a long, spiniform appendage (these spines very asymmetrical, too, and, generally speaking, directed posteriad: see Fig. 85). Inferior appendages bilobed in lateral view, superior lobe as an elongated and rather narrow, oblique band, much longer than the small, obtuse, inferior lobe; in dorsal view, the

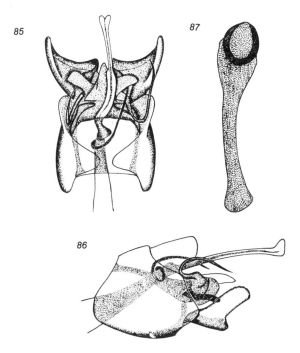

Figs. 85–87: *Agraylea* (*Allotrichia*) *vilnensis orientalis* ssp. nov.
85–86. male genitalia, specimen from Iran
(85. dorsal view; 86. lateral view);
87. longitudinal keel on sternite VIII of female, ventral view
(holotype from Lebanon).
(85–86: from Schmid, 1959)

inferior appendage appears rather short, with outer (= superior) lobe much more protruding than inner (= inferior) lobe, which is internally provided with (at least) one blunt, black "tooth"; the two lobes are widely separated by a rather shallow emargination. Phallus long, tubular, apex broadened and upturned; paramere very short.

Female genitalia: On sternite VIII, a long, sclerotized keel, almost reaching the end of this sternite; it is rather narrow (except for its two extremities), very distinctly capitate (the "head" distinctly framed with black), and uniformly coloured throughout its length. In the Iranian specimens (inf. Dr F. Schmid whose help is here acknowledged) it is exactly as I described it here.

Material: Female holotype from the Nabaa Joun spring, basin of the Nahr el Aouali (one of the small coastal basins of Lebanon), at 50 m a.s.l. Collected in daylight, by A. Dia, on 22 June 1980, and deposited in the Zoological Museum of the University of Amsterdam. Male allotype and 1 female paratype, presently not labelled as such:

same locality, same collector, collected on the same date as the holotype; these 2 specimens were at the University of Beirut, and it is unknown if they still exist.

Remarks: *Allotrichia vilnensis* was well described and illustrated by Raciecka (1937) from the vicinity of Vilna (presently Vilnius, Lithuania, U.S.S.R.). The species was again mentioned and illustrated by Schmid (1959), from material collected in northern Iran; this author mentions a few differences in the male genitalia between his specimens and those on which the original description was based, but decides that they have no specific value. This is probably true, but — see below — they certainly have subspecific value. The Lebanese specimens mentioned here are practically identical with those from Iran. *A. vilnensis* was also mentioned from Asia Minor (Malicky & Sipahiler, 1983), and from Latvia. The following characters (♂ genitalia) of *A. (A.) vilnensis vilnensis* (Raciecka, 1937) distinguish the nominate form from the new subspecies: gonopods with the two lobes more strongly developed, the emargination between them deeper — making the gonopod appear even shorter; two, instead of one, black "teeth" on the internal side of the median lobe (but this is not a really good distinctive character, the situation in the new subspecies being imperfectly known); the two lateral lobes of segment X, as well as their spiniform appendages, quite symmetrical. Strangely enough, in her (good) description of the female genitalia, Raciecka did not mention the sclerotized keel on sternite VIII — a conspicuous character, indeed. There is another strong argument pointing to *orientalis* as a distinct geographical race: *A. vilnensis* was never mentioned from the huge territories separating Lithuania and Latvia from the Near East: Poland, Russia, the Ukraine, the Caucasus.

The presence in the hindwings of *A. vilnensis* from Iran of furca 1, as emphasized by Schmid, should not be considered as an important character distinguishing it from the Lebanese specimens: the character is variable, and, e.g., in *A. vilnensis vilnensis*, this furca is either present, or absent.

A. (A.) vilnensis is related to several other species having inferior appendages resembling broad plates more or less deeply excised distally, and with black "teeth" on their internal side; one of these is *A. pallicornis* Eaton, whose presence in the Levant is to be expected.

Distribution and Ecological Notes: Moderately well distributed in northern Iran, the subspecies seems to be only quite sporadically distributed in Lebanon; its presence in Anatolia is possible. Moreover, *Allotrichia vilnensis* was recently mentioned from Bulgaria. The Lebanese specimens were caught in June, near one spring at 50 m a.s.l., but to term it a crenobiont would almost certainly be an error: data available from the Baltic (*A. vilnensis vilnensis*) and from Iran (*A. vilnensis orientalis*) show that the ecological spectrum of the species is a broad one, from springs to larger, muddy, lowland rivers.

56

Genus HYDROPTILA Dalman, 1819

K. svenska VetenskAkad. Handl., 40:125

Type Species: *Hydroptila tineoides* Dalman, 1819 (by monotypy).

Diagnosis: Comparatively large, very setose hydroptilids (Fig. 88), smaller than *Agraylea (Allotrichia)*, larger than the other genera present in the Levant, females larger than males, size of some species variable in different populations. The combination of the following characters makes the *Hydroptila* species unmistakable: ocelli lacking, spur formula 0, 2, 4, presence on the head (in the ♂ only) of dorsal postoccipital scent organs covered by mobile "caps" (these complex organs, with androconial setae and scales and producing pheromones, are species-specific, and their careful study could lead to fine results). Antennae even shorter in the female than in the male, with some 30 (or more, seldom less) segments in the male, with less than 30 in the female (interpopulational and individual variability); some groupings of segments with paler setae contrast with the darker vestiture of the remaining segments: this could be species-specific. Wings extremely setose, with numerous thickened, erect setae, forming a patchy pattern; they are somewhat narrower and more acuminate than in *Agraylea (Allotrichia)*, but clearly less so than in some other

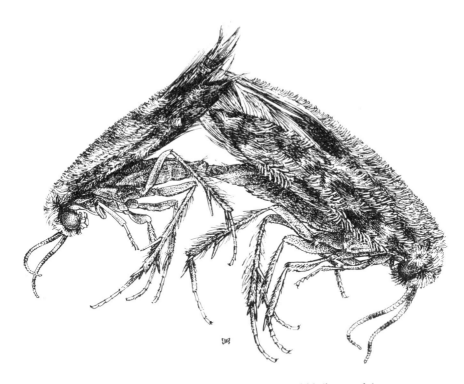

Fig. 88: *Hydroptila angustata* Mosely, 1939 (in copula)

57

Levantine genera; venation indistinct, only moderately reduced. In the male, on abdominal segment V there is a pair of minute latero-ventral setose tubercles, and on sternite VII a median process, usually long; on sternite VI in the female is a short, median process. The male genitalia are, generally speaking, symmetrical, but with asymmetrical structures too; segment VIII only seldom slightly modified; segment IX a well-developed sclerotized capsule, sometimes projecting anteriorly, almost always with postero-lateral projections; tergite X forming a "dorsal plate"; ventral parts of segment X of very varied shapes (see below); inferior appendages well developed, but usually simple, straight (only in the *vectis*-group with complex structure). Female genitalia forming a simple ovipositor, segment VIII short, modified, with diversely shaped "ventral sclerites", which possibly always correspond to internal glandular organs; there is always an internal trident-like structure in connection with the vagina; segment IX long, telescoped into the preceding ones; segment X short.

Represented in the Levant are 4 species-groups, characterized by features of the male and female genitalia: in the male the structure of the ventral parts of segment X, in the female the shape of the sclerites on sternite VIII . In the *sparsa*-group, a very large Palaearctic group: male without conspicuous appendages, with a very simple "subgenital plate"; female with a mushroom-like structure. In the *occulta*-group, a large group with Palaearctic, Ethiopian, Oriental, and Nearctic species: male with a pair of very diversely shaped, long, often curiously asymmetrical appendages, the "ventral branches of segment X"; female with a tribranched structure. In the *pulchricornis*-group, a very small, mainly Palaearctic group: male with an obliquely down- and backwards directed structure framed by sclerotized stripes converging to a common "head"; female without conspicuous sclerites. In the *vectis*-group, consisting of two Palaearctic species: male with one or two long, slender, black spines; female with longitudinal median sclerotized structure.

General Remarks: *Hydroptila* is a very large genus. In Marshall (1979) some 150 species are listed (a few were omitted), but presently, possibly some 60 more are known, and every year additional ones are described. This is the Trichoptera genus with the largest number of species in the Levant, where, I suspect, some more species will be found, especially in the northernmost parts. The genus is cosmopolitan, with almost world-wide distribution. There is no comprehensive revision of the genus; much interesting information will certainly become available when its phylogeny is worked-out on a biogeographic background.

The larvae and pupae, morphologically very similar, build laterally compressed cases, resembling spectacle-cases, of secretion mostly covered with fine sand grains. In the Levant no species was found in stagnant water, but all types of running water are inhabited, from the Crenal of higher mountains to the Epipotamal of the lowlands and to the desert watercourses. Of course, each species has its own ecological spectrum, and all the intermediate situations are found between apparently stenobiont species like *H. fonsorontina* or *H. viganoi*, and the most eurybiont ones, like *H. angustata* or *H. aegyptia*. Several species are important members of the communities of different types of running water.

Hydroptilidae: Hydroptila

Key to the Species of Hydroptila in the Levant

♂♂

1. From below the dorsal plate of segment X, one or two, long, slender, black spines arise; inferior appendages with dorsal and ventral branch forming a clasping tool

2 (gr. *vectis*)

– No such black spines; inferior appendages simple, not branched 3

2. One pair of symmetrical black spines; ventral branch of inferior appendages distinctly bifid. **H. vectis** Curtis

– Only one black spine, on the right side; ventral branch of inferior appendages not bifid.

H. viganoi Botosaneanu

3. Sternite VIII deeply excised posteriorly; ventral parts of segment X represented by an oblique "plate" framed by dark, sclerotized stripes, capitate apically where it is produced in a beak directed downwards. **H. aegyptia** Ulmer

– Sternite VIII not excised; ventral parts of segment X having other shapes 4

4. Ventral parts of segment X a simple, inconspicuous "subgenital plate", seen between the inferior appendages 5 (gr. *sparsa*)

– Ventral parts of segment X represented by a pair of long appendages variously conformed

7 (gr. *occulta*)

5. Dorsal plate (segment X) with apical angles distinctly indented; inferior appendages in lateral view with sharply produced and upturned upper angle of apex, and with black "tooth" near inferior margin widely distant from apex.

H. phoeniciae Botosaneanu & Dia

– Dorsal plate with apical angles without indentations; inferior appendages, in lateral view, with less, or much less, sharply produced and upturned upper angle of apex, and black "tooth" near inferior margin either lacking or slightly distant from apex 6

6. Inferior appendages very slender, in lateral view not capitate, very slightly upturned at apex, without black "tooth" near inferior margin. **H. sparsa** Curtis

– Inferior appendages, in lateral view, distinctly capitate, upper apical angle upturned like a strong beak, near ventral margin a black "tooth" subapically placed and united to the upper angle by a black line. **H. angustata** Mosely

7. "Ventral branches of segment X" not coalescent basally, subapically crossing each other, or also strongly turned cephalad 8

– "Ventral branches of segment X" variously shaped but arising from a common root, subapically not crossing each other, not strongly turned cephalad 9

8. Arising from the posterior margin of sternite IX is a long, strong appendage terminating in a hook; inferior appendages (lateral) basally narrow, then strongly enlarged, with blunt apices; "ventral branches of segment X" characteristically crossing each other subapically, not directed cephalad. **H. hirra** Mosely

– No such appendage arising from the margin of sternite IX; inferior appendages (lateral), very broad (= high) basally, then suddenly narrowing to slender appendages apically hooked; "ventral branches of segment X" in their subapical parts strongly turned cephalad, above the dorsal plate. **H. adana** Mosely

9. "Ventral branches of segment X" almost symmetrical, not branched.

H. mendli levanti Botosaneanu

– "Ventral branches of segment X" strongly asymmetrical, left branch always splitting into two rami, right branch shorter, not split 10

59

10. (a) "Ventral branches of segment X" with right branch much shorter than left branch and ending in a strong hook directed dextrally.   **H. palaestinae** Botosaneanu & Gasith

(b) "Ventral branches of segment X" with right branch much shorter than left branch and ending in a simple, minute cone.   **H. libanica** Botosaneanu & Dia

(c) "Ventral branches of segment X" with right branch shorter than left branch, but nevertheless ending in an appendage resembling the rami of the left branch.

H. fonsorontina Botosaneanu & Moubayed

♀♀

(unknown for *H. mendli* and *H. viganoi*)

1. No sclerites on abdominal sternite VIII, this sternite much longer than the respective tergite, both deeply bilobed.   **H. aegyptia** Ulmer

– Conspicuous sclerotized formations on (in) sternite VIII   2

2. On sternite VIII, a longitudinal median sclerotization; lateral branches of the trident-like internal structure extremely long and slender.   **H. vectis** Curtis

– Sclerotized formation on sternite VIII either mushroom-like or tribranched   3

3. Sclerotized formation on sternite VIII mushroom-like   4 (gr. *sparsa*)

– Sclerotized formation on (in) sternite VIII tribranched   5 (gr. *occulta*)

4. (a) "Mushroom" with distal part transversely-ovoid, not (= very slightly) produced on the sides; internal "trident" with broad stalk which is longer than the lateral arms.

H. sparsa Curtis

(b) "Mushroom" with distal part slightly angular and flattened, distinctly produced on the sides; internal "trident" with short, basally usually truncate stalk (shorter than or as long as the lateral arms).   **H. angustata** Mosely

(c) "Mushroom" with angular, flattened but high apical part, rather well produced on the sides; internal "trident" with long and relatively narrow stalk, as long as or slightly longer than the lateral arms.   **H. phoeniciae** Botosaneanu & Dia

5. On distal margin of sternite VIII, two groups of strong setae which are widely separated on the median line; tribranched sclerotized formation with short lateral branches, somewhat swollen central branch resembling a rounded, transversal, reticulated shield.

H. hirra Mosely

– The strong setae on distal margin of sternite VIII not separated into two groups   6

6. The 6 strong setae distally on sternite VIII forming a concave or straight row (not convex); tribranched sclerotized formation with long, swollen, lateral branches, central branch resembling a quadrangular, somewhat longitudinally extended shield.

H. adana Mosely

– The 6 strong setae distally on sternite VIII in a convex row   7

7. Distal, convex margin of sternite VIII, on which the 6 strong setae are inserted, strongly sinuous; tribranched sclerotized formation with the three branches swollen, central part well distinct from the rest; there is a conspicuous dark "hump" on each side of this formation.   **H. libanica** Botosaneanu & Dia

– Distal, convex margin of sternite VIII not strongly sinuous   8

8. Tribranched sclerotized formation with branches not swollen, Y-shaped

H. palaestinae Botosaneanu & Gasith

– Tribranched sclerotized formation with branches distinctly swollen.

H. fonsorontina Botosaneanu & Moubayed

Hydroptila sparsa Curtis, 1834

Figs. 89–94

Hydroptila sparsa Curtis, 1834, *Phil. Mag.*, 4:217.

In an important sample of males from Baalbek, Lebanon, the wing expanse varies from 5.5 to 7.4 mm, that of most of the specimens being 6.5–7 mm; only a few females could be accurately measured, their wing expanse being 6.2–7.7 mm. Antennae (♂) with 34–38 segments, most of the specimens having 36–38 segments. Wings: Fig. 89.

Male genitalia: segment IX globose, in lateral view moderately high at its distal end, with very obtuse proximal end, latero-distal projections long (more than half the length of the gonopod), conical, pointed. Dorsal plate formed by segment X with rather small but very distinct triangular medio-apical indentation, apical corners sharply produced. Ventral, or "subgenital" plate (seen between the inferior appendages) with not very deep distal emargination and blunt angles. Inferior appendages of simple shape, in lateral view slender, only with a minute hook at apex (directed dorsad and laterad), in ventral view strongly divergent, gradually narrowing towards the tips, without distinct black "teeth". Apex of phallus with hook directed dextrally.

Female genitalia: Distal part of the "mushroom" (sclerotized structure of sternite VIII) transversely ovoid, not produced on the sides. Trident-like internal structure with rather long and distinctly broad stalk, longer than lateral arms, lateral borders of this stalk sinuous (but the shape of the trident-like structure seems to be subject to rather important variation!).

Remarks: *H. sparsa* belongs to the species-group bearing the species' name. There is apparently no important variability in the genitalia, but an interesting geographic variability is noted in the number of antennal segments: in populations from western and eastern Europe the number mentioned by practically all the authors is 32 (or "ca. 32") in the male, in clear contrast with the situation in the Lebanese populations.

Distribution and Ecological Notes: The species is known from almost every zone of Europe, except the northernmost ones, as well as from Asia Minor and from northern Iran; in Europe it is an inhabitant of the Potamal, being found mostly at low altitudes. In Lebanon it was never found in the coastal hydrographic systems studied up to the present, but is abundant in the hydrographic systems of the Orontes and of the Litani, being found (sometimes in large numbers) in springs — especially large karstic springs — and in smaller and larger streams, as well as in the main watercourses (10–15 m broad), at least between ca. 600 and ca. 1,200 m a.s.l.; thus here it is an inhabitant of the Crenal and of all zones of the Rhithral. In Israel, *H. sparsa* is known only from the spring complex of Tel Dan (Tel el Kadi) at 200 m a.s.l. (1), and even here the population is certainly not very numerous, pointing to the fact that this is probably the southernmost point in the species' distribution. We have in this species a fine illustration of the principle of ecological vicariance. In the Levant *H. sparsa* is on the wing at least from March to October; in the samples males always exceed the females in numbers.

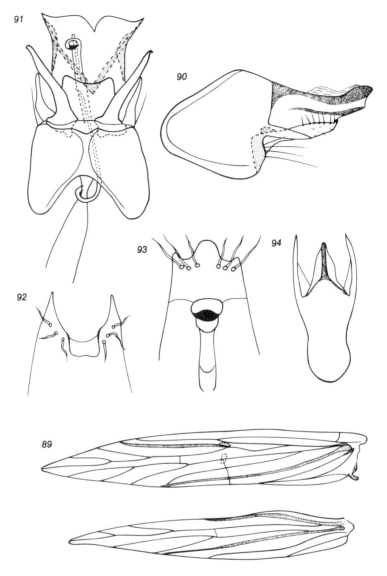

Figs. 89–94: *Hydroptila sparsa* Curtis, 1834
89. wings;
90–91. male genitalia
(90. lateral view; 91. ventral view);
92–94. female genitalia
(92. segment VIII, dorsal view, 93. ventral view;
94. trident-like internal structure, ventral view)

Hydroptila angustata Mosely, 1939

Figs. 88, 95–100

Hydroptila angustata Mosely, 1939b, *Ann. Mag. nat. Hist.*, (11)3:43–48.
Hydroptila simulans Mosely, 1919. (auct. e.g., Botosaneanu & Gasith, 1971, *Israel J. Zool.*, 20:98).
Hydroptila neglecta Kumanski, 1983, *Reichenbachia*, 21(2):15–18 (synonymized by K. Kumanski, 1985:124).

Wing expanse: In the Levantine specimens — ♂ 4.3–5.5 mm (ca. 4.3 mm in the type from Egypt), ♀ 5.6–6.2 mm. Antennae in the ♂ with 30–32 segments (only 29 in the type from Egypt); in the unique presently examined female specimen from Israel with intact antennae, 26 or 28 segments. Habitus: Fig. 88.

Male genitalia: The figures given by Mosely (1939b) in his original description are reproduced here, because, despite minor imperfections, they are representative of the Levantine specimens too (the exact shape of segment IX is not depicted in these figures — but this is not very important; Figs. 95 and 96 were slightly improved by me). Distal margin of tergite IX roundly produced; latero-distal projections of segment IX acute, not very long. The dorsal plate formed by segment X is long, with a deep triangular excision of the apical margin, starting from the apical angles, which are rather acute. The ventral, or subgenital plate, seen between the inferior appendages ("lower penis-cover"), is apically emarginated, its lateral angles broadly rounded. Inferior appendages, in lateral view, broader at the base, then very slender in the middle, with a distinctly broader apical part having a bluntly rounded lower angle, the upper angle being produced in a strong, sharp "beak" directed upwards; this beak is connected with the lower margin of the appendage (where a minute black wart is present) through an oblique keel accompanied by setae; in ventral view, the inferior appendages are conical, divergent. At apex of phallus, a strong hook curled dextrally.

Female genitalia: Mushroom-like sclerotized structure of sternite VIII, with apical part slightly angular and flattened, its lateral angles very well produced. Trident-like internal structure with very short stalk (shorter than, or as long as, the lateral arms of the trident), proximally more or less truncated (sometimes rounded); the central branch of the trident with a distinct pair of lateral "wings".

Remarks: *H. angustata* belongs to the *sparsa* species-group, like *H. sparsa* and *H. phoeniciae*. It was confused several times with the closely related *H. simulans* Mosely, 1919.

Distribution: This species was originally described from the oasis of Zeitoun, in Egypt. It is presently clear that it has a wide distribution all around the Mediterranean — at times penetrating rather deep inside southern Europe or the Near East — from Spain to Iran (and possibly Asia Minor), from Egypt to Bulgaria.

In Lebanon several localities are known in the small coastal hydrographic systems: Nahr el Aouali with various localities between 50 and 380 m a.s.l., the lower course of Nahr ed Damour, and also the Nabaa Joun spring at 36 m a.s.l.; the species was also caught on the Litani at Jib-Jennine (800 m a.s.l.) as well as at 200 m a.s.l., in the

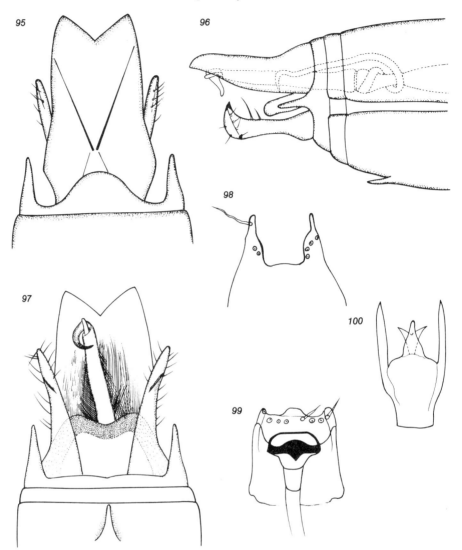

Figs. 95–100: *Hydroptila angustata* Mosely, 1939
95–97. male genitalia, specimen from Egypt
(95. dorsal view; 96. lateral view; 97. ventral view);
98–100. female genitalia
(98. segment VIII, dorsal view; 99. segment VIII, ventral view;
100. trident-like internal structure).
(95–97: from Mosely, 1939b; 95–96 slightly modified)

lower course of this river, 15 km from its mouth. It will certainly be found in many other Lebanese localities. The known localities in Israel are rather numerous (only some of them previously published); it can be said that *H. angustata* is presently known from a vast area in eastern Israel, from the Galilee to the 'Arava Valley; the localities are: Tel Dan (1) and 'En Tut (3), Deganya B (possibly the Jordan itself), HaMovil HaMaluach (Tiberias), and Bet She'an (7), the desert streams Naḥal 'Arugot and Naḥal Dawid, both at 'En Gedi, the 'Ein Nueima spring (13), and Ne'ot haKikar (14). These localities are between 200 m a.s.l. and 350 m below sea level. It is somewhat surprising that it was never caught in the Negev proper; it is probably present in the Judean Hills.

Ecological notes: This *Hydroptila* species is one of the most ubiquitous, tolerant of environmental factors, probably opportunistic and expansive species. It inhabits very different watercourses (from the Crenal to the Epipotamal), at lower altitudes (from 350 m below sea level to ca. 800 m a.s.l.), being mostly present in slowly flowing water with fine particulate substrates and high and strongly variable temperatures. It does not avoid waters with rather high saline concentrations (e.g., the spring-fed waters at HaMovil HaMaluach, or the spring at Ne'ot HaKikar). The species is on the wing throughout the year. According to Gasith (1969), it completes one generation in only 1.5–2 months.

Hydroptila phoeniciae Botosaneanu & Dia, 1983

Figs. 101–107

Hydroptila phoeniciae Botosaneanu & Dia, in: Dia & Botosaneanu, 1983, *Bull. zool. Mus. Amsterdam*, 9(14):125–135. Type Locality: Nabaa Bâter ech Chouf, downstream from village Niha, Lebanon. Type deposited in the Zoological Museum, University of Amsterdam.

Wing expanse: ♂ 5.5–6.5 mm. Antennae (♂) with 35 segments.
Male genitalia: Segment IX globose, very obtuse anteriorly, with a dorso-median zone distinct from the rest, with latero-posterior projections rather long, obliquely directed ventrad, pointed. Segment X represented by a very characteristically shaped plate; in dorsal view this is broad proximally, then suddenly narrowed, with distal part again broadened; apically the plate has a very deep, triangular sinus, and each of the two apical angles is clearly indented. Rather deeply emarginated "subgenital plate". Inferior appendages not very strongly broadened towards the apex, apical superior angle produced in a strong point nearly vertically directed upwards (and also laterally); the strong black "wart", laterally placed near the inferior edge of the appendage, is widely distant from the apical lower angle, and is not connected through a keel (or line) to the superior edge. Phallus ending in a hook curled sinistrally; its part distad from the root of the paramere is remarkably long.
Female genitalia: Mushroom-like sclerotized structure of the sternite VIII with slightly angular and high apical part, lateral angles moderately well produced. Trident-like internal structure with a long and narrow stalk, slightly longer than or as long as the lateral arms; proximal end rounded.

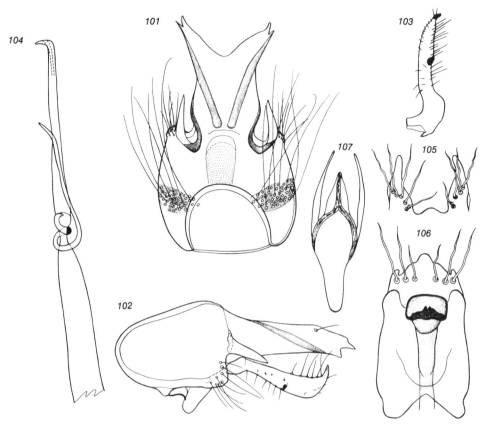

Figs. 101–107: *Hydroptila phoeniciae* Botosaneanu & Dia, 1983
101–104. male genitalia
(101. dorsal view, the two dark spots with long hairs
belong to tergite VIII, not tergite IX; 102. lateral view;
103. right inferior appendage, ventral view; 104. phallic apparatus, ventral view);
105–107. female genitalia
(105. segment VIII, dorsal view; 106. segment VIII, ventral view;
107. trident-like structure).
(From Dia & Botosaneanu, 1983, except 107)

Remarks: *H. phoeniciae* belongs to the *sparsa* species-group, like the two preceding species, but is possibly not very closely related to them; in certain male genital characters (perhaps misleading) it somewhat resembles *H. emarginata* Mart. (Central Asia, Iran, N. Korea). I feel obliged to digress here. *H. bajgirana* Bots., 1983, is undoubtedly a synonym of *emarginata*, which is, in my opinion, a valid species. I disagree with Malicky (1986b:6) who does not consider *H. emarginata* Mart., 1927, as being distinct from the — admittedly closely related — *H. angulata* Mos., 1922: *emarginata* can always be distinguished from *angulata* by the clearly indented apical angles of the dorsal plate (segment X) in the male, a character shared with *H. phoeniciae*. Having examined a large series of male *H. emarginata* from N. Korea,

Dr K. Kumanski (Sofia; in lit.) agrees with my idea that *H. emarginata* and *H. angulata* are clearly distinct.

Distribution and Ecological Notes: Known only from several localities in two small coastal basins of Lebanon, Nahr el Aouali and Nahr ed Damour, where it inhabits springs and smaller and larger streams, including the main watercourses, from very low locations to at least 850 m a.s.l. Found on the wing in April–June and in November.

Hydroptila mendli levanti Botosaneanu, 1984
Figs. 108–115, 148

Hydroptila mendli levanti Botosaneanu, 1984b, *Ent. Ber., Amst.*, 44:137–139. Type Locality: Nahr ed Damour downstream confluence with Nahr el Hammam, Central Lebanon. Type deposited in the Zoological Museum, University of Amsterdam.

Description of ♂ (♀ presently unknown): Wing expanse about 4.7–4.9 mm. Antennae mostly with 29 segments, sometimes with 30. Segment IX moderately produced anteriad; proximal end obtuse; dorsal edge, in lateral view, gradually sloping down anteriad; latero-posterior projections ogival, strong. Dorsal part of segment X represented by an elongated plate, clearly broader (more swollen) in its middle than at the two ends; at its distal margin 6 small projections, or lobes, can be seen, separated by 5 small sinuses (the two median lobes membranous, the 2 pairs of lateral lobes sclerotized). "Ventral branches of segment X" simple, strongly upturned in their distal half, symmetrical in their proximal parts, but very slightly asymmetrical in the distal ones: left branch longer and more slender then the right one, and with honeycombed apex slightly less developed. The inferior appendage is an elongated, narrow band of irregular shape; it runs first straight backwards, but at the beginning of its distal part there is a very slight hump on the dorsal side, and then the gonopod turns slightly obliquely downwards to the rounded tip; in ventral view (shape strongly influenced by the observation angle, this being true also for the lateral shape!), distal part of the gonopod very slightly bilobed, one of the lobes corresponding to the tip, the other to the already mentioned hump.

Remarks: *H. mendli levanti* belongs to the *occulta* species-group. It was described as a subspecies of *H. mendli* Malicky, 1980, first described from the High Atlas, Morocco (Malicky, 1980), and subsequently discovered also in the Middle Atlas. The differences between what seem to be two geographical races are extremely slight, and further study, including that of specimens possibly to be discovered in localities between the Atlas and Lebanon, will perhaps show that the Lebanese populations do not merit even the subspecific rank. The characters distinguishing *H. mendli levanti* from the nominate form are: size slightly larger, antennae with very slightly larger number of segments, and especially the slightly asymetrical "ventral branches of segment X".

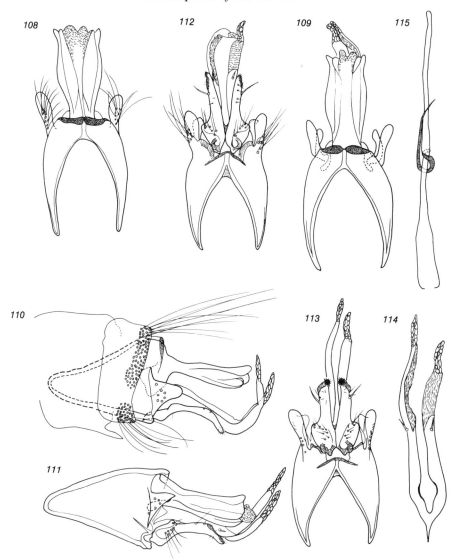

Figs. 108–115: *Hydroptila mendli levanti* Botosaneanu, 1984, male genitalia,
two different Lebanese specimens
(108–109. dorsal view; 110–111. lateral view; 112–113. ventral view;
114. "ventral branches of segment X", ventral view;
115. phallic apparatus, dorsal view).
(From Botosaneanu, 1984b)

Distribution and Ecological Notes: Presently known only from a few specimens collected in July and September on the final 4 km of the small river Nahr ed Damour, in Central Lebanon, at very low altitudes.

Hydroptila palaestinae Botosaneanu & Gasith, 1971

Figs. 116–121, 149

Hydroptila palaestinae Botosaneanu & Gasith, 1971, *Israel J. Zool.*, 20:99–102. Type Locality: Bet She'an, Israel. Type deposited in the Zoological Museum, Tel Aviv University.

Wing expanse: About 4.5 mm in both sexes. Antennae of ♂ with 30 segments, those of ♀ with 22 segments.

Male genitalia: Segment IX deeply excised dorsally and ventrally, dorsal excision wider than the ventral one; in lateral view it appears moderately stout, with obtuse anterior end and rather short but broad triangular latero-posterior projections; but in the specimen from "Tel el Kadi?", mentioned below, it is distinctly longer and more slender, dorsal margin not convex as in Fig. 117, but slightly concave, and the latero-posterior projections are broadly rounded at apex. Plate of segment X with sinuous margins, narrowest at base and broadest at apex; its central part is membranous, bilobed at apex, and bordered by two sclerotized lateral stripes, each of them deeply emarginated (bilobed) at apex. "Ventral branches of segment X" very characteristically shaped: their "root" is symmetrical, but the branches are completely asymmetrical, with the right branch much shorter than the left, obliquely directed downwards and sinistrally, ending in a strong hook directed dextrally; left branch much longer, in its proximal, baculiform part directed obliquely downwards, then abruptly twisted at right angles, and at the same time splitting into two ramifications directed obliquely upwards and backwards (Fig. 117); these 2 ramifications are asymmetrical: the shorter one is ventral and on the right side, the longer one is dorsal and on the left side (Figs. 117, 149), both of them swollen in their proximal parts, then strongly narrowed in their honeycombed apical parts. Inferior appendages completely coalescent in their basal half (arising from a common basal plate); in dorsal and ventral view they are moderately narrow, divergent stripes, apical edge obliquely cut, blackened, slightly emarginated (viewed at certain angles, the apex can resemble a distinct hump); in lateral view, they are very narrow at base, then broader, dorsally swollen, and directed slightly downwards (shape varying rather markedly with the observation angle !).

Female genitalia: Terminally on the ventral side of abdominal segment VIII, a semicircular crown of 6 strong setae; the most conspicuous structure to be seen inside sternite VIII is a very simple Y-shaped structure, representing a glandular organ with sclerotized walls.

Remarks: Belonging to the *occulta* species-group, *H. palaestinae* is very closely related to *H. libanica* and *H. fonsorontina* (see below), these 3 species forming a distinct sub-group, possibly with Oriental affinities. In the above description, the existence of a curious variability in the shape of segment IX (perhaps characteristic of two populations?) was already mentioned.

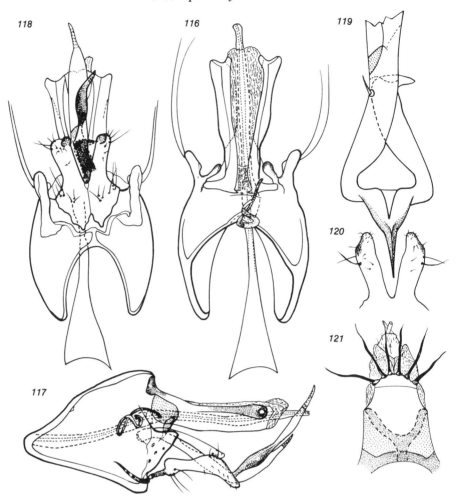

Figs. 116–121: *Hydroptila palaestinae* Botosaneanu & Gasith, 1971
116–120. male genitalia
(116. dorsal view; 117. lateral view, 118. ventral view;
119. basal part of "ventral branches of segment X",
only with right branch entirely represented; 120. inferior appendage, ventral view);
121. female genitalia, ventral view.
(From Botosaneanu & Gasith, 1971)

Distribution and Ecological Notes: Known from Bet She'an, in the middle part of the Jordan River Valley, Israel (7). Here, numerous specimens, pointing to an apparently flourishing population, were caught in February 1968, along the "-200 canal" which is an aqueduct with fast flowing, rather turbid, and eurythermic water originating from a spring ('En 'Amal) about 6 km westwards from the collecting site. There is also 1 male with morphological peculiarities of segment IX, labelled "Tel el Kadi? I.67".

Hydroptila libanica Botosaneanu & Dia, 1983
Figs. 122–127, 150

Hydroptila libanica Botosaneanu & Dia, in: Dia & Botosaneanu, 1983, *Bull. zool. Mus. Amsterdam*, 9(14): 130–131. Type Locality: Nahr ed Damour, downstream from confluence with Nahr el Hamman, Central Lebanon. Type deposited in the Zoological Museum, University of Amsterdam.

Wing expanse (♂): 4.5–5 mm. The antennae of the ♂ holotype have 34 segments, those of a ♂ paratype only 32.

Male genitalia: Segment IX deeply excised ventrally, and even more so dorsally; in lateral view distinctly elongate and moderately slender, with moderately obtuse anterior end, and characteristically shaped latero-posterior projections, which are rather long, basally broad but then forming narrow stripes, apically rounded. Plate of segment X without really distinctive characters if compared, e.g., with that of *H. palaestinae*. The "ventral branches of segment X" are perfectly symmetrical in their basal parts, where the root of each branch resembles a strong "hook" pointed mediad; the branches are markedly asymmetrical; the right branch remains short, reaching only the forking point of the left branch; there is a subapical small spine, and the apex of this (right) branch is merely a very small conical projection, without any kind of appendage; the left branch further splits into two highly asymmetrical strong appendages: the shorter one is dorsally placed, directed obliquely sinistrally, upwards and backwards; the longer one has a slightly more ventral position, runs obliquely upwards and backwards, but in the middle of its length it is suddenly twisted at right angles and extends dextrally; terminal parts of the two appendages honeycombed. Inferior appendages, in ventral view, divergent, completely separated on the median line (until the base); apex more or less distinctly bilobed, the two lobes rounded, of equal size, blackened; lateral shape not very original, dorsal swelling well developed.

Female genitalia: The association of the female of this species is slightly uncertain. However, since many females were caught at one locality together with a male paratype, and no other species of the *occulta* group was present at that locality, these specimens may be considered as being almost surely *H. libanica*. Terminally on the ventral side of abdominal segment VIII, a semicircular crown of strong setae inserted on a strongly sinuate margin. The "glandular organ" internal to sternite VIII has its 3 branches swollen, its central part well distinct from the rest. On both sides of this organ, segment VIII has a conspicuous, dark "hump", and there are several transverse "humps" medially, anteriad to the "glandular organ".

Remarks: Belonging to the *occulta* species-group, *H. libanica* is very closely related to *H. palaestinae* and especially to *H. fonsorontina*, these 3 species forming a distinct subgroup, possibly with Oriental affinities.

Distribution and Ecological Notes: Rather many specimens of *H. libanica* were caught in the basin of the small Lebanese coastal river Nahr ed Damour; this species was found here in the main watercourse, at 260 m a.s.l., but also in its lower reaches

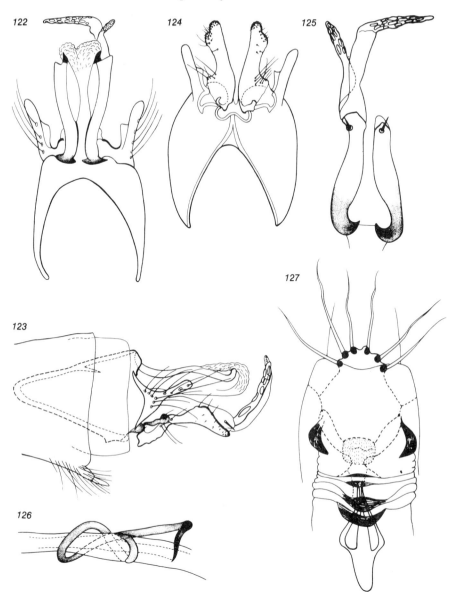

Figs. 122–127: *Hydroptila libanica* Botosaneanu & Dia, 1983
122–126. male genitalia
(122. segments IX and X, dorsal view; 123. lateral view;
124. segment IX and gonopods, ventral view;
125. "ventral branches of segment X", ventral view;
126. central zone of phallic apparatus with titillator, lateral view);
127. female genitalia, essentially segment VIII, ventral view.
(From Dia & Botosaneanu, 1983, except 127)

(0–40 m a.s.l.), and on a tributary, Nahr el Hamman, at 45 m a.s.l. I have also seen 2 males which were caught on Nahal Senir (Hazbani), one of the main branches of the Upper Jordan River, in Israel (1), this being probably the southernmost point in the species distribution. It will be noted that in all these cases we are concerned with larger watercourses, belonging probably to the Metarhithral and Hyporhithral. The Lebanese specimens were caught in July–August; the specimens from Israel (Hazbani) — in April and October respectively.

Hydroptila fonsorontina Botosaneanu & Moubayed, 1985
Figs. 128–130, 151

Hydroptila fonsorontina Botosaneanu & Moubayed, in: Moubayed & Botosaneanu, 1985, *Bull. zool. Mus. Amsterdam*, 10(11):64–65, Figs. 2–4. Type Locality: the Orontes at Hermel, downstream from the Zarka spring, Lebanon. Type deposited in the Zoological Museum of the University of Amsterdam.

Wing expanse: ♂ type, 5.3 mm; ♀♀ 3.7–5 mm. Antennae of ♂ with 32–33 segments, of ♀ with 25 segments.
Male genitalia: Generally very similar to those of *H. libanica*, and only the distinguishing characters will be mentioned here. In lateral view, the latero-posterior projections of segment IX are large, simply triangular. "Ventral branches of segment X": left one branching off in the same manner as in *libanica*, and the relations between the two resulting appendages are similar, too, but the longer (ventral) appendage is never twisted at right angles, its terminal, honeycombed part being very obliquely directed backwards. However, the main distinguishing character is offered by the right, uniramous branch: in *H. fonsorontina* it extends far beyond what we find, in *H. libanica*, as being a small subapical spine: this terminal part of the branch is first membranous, then sclerotized and honeycombed, twisted to the right side (in lateral view it is twisted almost at right angles).
Female genitalia: Terminally on the ventral side of abdominal segment VIII, a semicircular crown of strong setae. The "glandular organ" within sternite VIII has all its three branches distinctly swollen.
Remarks: Belonging to the *occulta* species-group, *H. fonsorontina* is closely related to *H. palaestinae*, and certainly a sister- species of *H. libanica*. Despite the great general similarity of *H. fonsorontina* and *H. libanica*, the distinguishing characters (♂ genitalia) here evoked are absolutely constant in the large number of examined specimens.
Distribution and Ecological Notes: Presently known only from the type locality: the Orontes River downstream from one of its main springs (Zarka), at Hermel, 650 m a.s.l. A large population develops in this fast flowing, mainly stony stream.

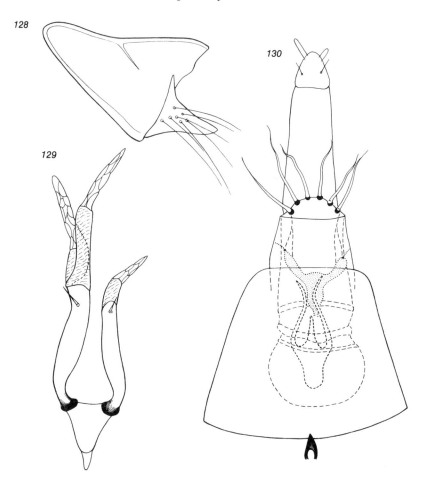

Figs. 128–130: *Hydroptila fonsorontina* Botosaneanu & Moubayed, 1985
128. segment IX of male, lateral view;
129. "ventral branches of segment X" of male , ventral view;
130. female genitalia, ventral view.
(From Moubayed & Botosaneanu, 1985)

Hydroptila adana Mosely, 1948

Figs. 131–133, 135–143, 152

Hydroptila adana Mosely, 1948, *British Museum Expedition to South-West Arabia 1937–8*,
1:81, Figs. 33–39. Botosaneanu & Gasith, 1971, *Israel J. Zool.*, 20:99, Fig. 2; Botosaneanu,
1973, *Fragm. ent.*, 9:61–80, Fig. 2.

Wing expanse: ♂ 5.1–5.3 mm, ♀ 5.7–6.7 mm. Antennae in the ♂ always with 32
segments (31 were found by Mosely in specimens from Yemen), in the ♀ always with

25 segments. Wing venation: Fig. 131. The scent organs on the head of the ♂ have conspicuously developed caps (Figs. 132–133), and one of their constituent parts is a dense brush of long, black, converging setae arising from a blackened, circular plate.

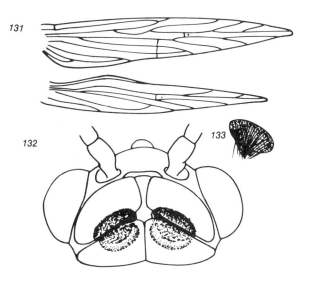

Figs. 131–133: *Hydroptila adana* Mosely, 1948
131. wings; 132. head with scent-organs; 133. "brush" of a scent-organ
(From Mosely, 1948)

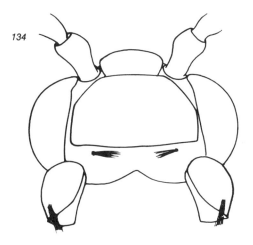

Fig. 134: *Hydroptila hirra* Mosely, 1948, head with scent-organs (hanging below).
(From Mosely, 1948)

Male genitalia: Arising from abdominal tergite VIII, two compact and conspicuous brushes of extremely long, black setae. Segment IX elongate in lateral view, very high distally but strongly tapering towards the anterior end, with long latero-posterior projections with almost parallel sides and blunt apex; posterior margin of sternite IX with median triangular projection. Plate of segment X excised at apex in the particular manner shown in Fig. 138, its apical angles rounded and turned under. The "ventral branches of segment X" are practically symmetrical: their proximal, almost parallel, halves have a ventral position; then the appendages are very strongly curled up and then forwards over the dorsal plate of segment X; they normally cross over in their anteapical parts, and each of them has in this anteapical part a much smaller additional branch (apices of main and additional appendages very slightly honeycombed). Inferior appendages, when seen in lateral view, very high in basal parts (broad here in ventral view), then suddenly extending into long, slender, either parallel or diverging appendages tapering towards the apices which are blackened and very strongly turned laterad and also upwards. Phallus very long, slender, anteapically produced in a laterally and downwards directed "beak".

Female genitalia: Terminally on the ventral side of abdominal segment VIII, a row of 6 strong setae inserted on a concave margin (sometimes appearing as straight, horizontal, but never convex). The glandular organ within sternite VIII is extremely well sclerotized, dark, its long lateral branches swollen, its central branch strongly individualized, like a more or less quadrangular, longitudinally developed "shield"; laterally to this organ, segment VIII has a pair of "humps", but there seem to be no (important, at least) median "swellings" anterior to the glandular organ.

Remarks: *H. adana* belongs to the *occulta* group of species, but is certainly not closely related to the other species belonging to the Levantine fauna. Some characters of the male genitalia and especially of segment IX and of the "ventral branches of segment X" point to affinities with the Pakistani species *H. sengavi* Schmid (Schmid, 1960:93–94, Pls. 6–7) and, to a lesser degree, with other Pakistani species. It is a very specialized species, and such characters as the enormous tufts of setae on tergite VIII, the peculiar shape of the inferior appendages and of the "ventral branches of segment X" in the male, and the shape of the glandular organ and of the row of apical setae of sternite VIII in the female, will allow its recognition without difficulty.

The species shows no interesting variability. If the illustrations in Mosely (1948) and Botosaneanu (1973) are compared, there may appear to be some differences between male specimens from Yemen and from the Sinai and the Dead Sea Depression, but this is not true: the gonopods are either parallel or diverging (compare Figs. 138 and 139), and Mosely correctly reproduced the triangular median projection on the posterior edge of sternite IX (Fig. 139), which was omitted in Fig. 138.

Distribution and Ecological Notes: This species was described from Wadi Dhahr, north-west of Sana, Yemen, at ca. 2,600 m a.s.l. It was subsequently caught in the Sinai Peninsula at Wadi Isla, 700 m a.s.l. and at Wadi Watir, 350 m a.s.l. (22), and also in the Dead Sea Depression, at Nahal 'Arugot and at 'Ein Turabe, at ca. 400 m below sea level (13). No doubt *H. adana* will be discovered in other suitable localities in the

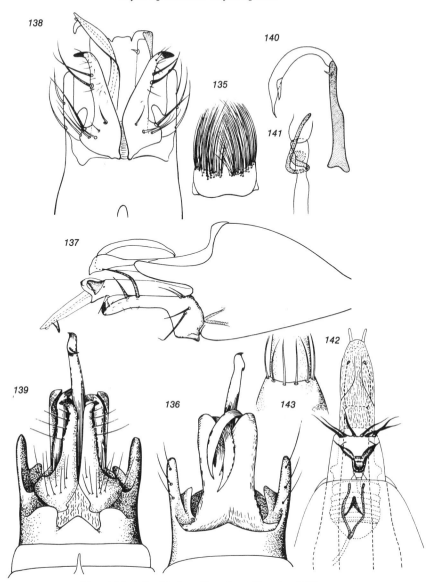

Figs. 135–143: *Hydroptila adana* Mosely, 1948
135–141. male genitalia
(135. tergite VIII; 136. dorsal view; 137. lateral view; 138–139. ventral view;
140. right branch of "ventral branches of segment X";
141. central zone of phallic apparatus with titillator);
142–143. female genitalia
(142. ventral view; 143. distal part of sternite VIII with its row of setae);
Figs. 136, 139, 143: specimens from Yemen.
(135, 137, 138, 140, 141, 142: from Botosaneanu, 1973;
142: from Botosaneanu & Gasith, 1971;
136, 139, 143: from Mosely, 1948)

Sinai and the Negev, but it is certainly a less frequent species than *H. hirra* (in whose company it is regularly found) and much less expansive than the latter: in the Levantine Province, *H. adana* was never found north of the northern limits of the Dead Sea Depression. One male of this species was mentioned (Malicky, 1986b:9) from Iran, 40 km north of Bandar Abbas.

The inhabited biotopes are always either springs, spring-fed streamlets or larger streams, in desert environments, and there is an impressive altitudinal range. Adults were caught in April, September, and December (January in Yemen): the species is possibly on the wing throughout the year.

Hydroptila hirra Mosely, 1948
Figs. 134, 144–147, 153

Hydroptila hirra Mosely, 1948, *British Museum Expedition to South-West Arabia 1937–8*, 1:81–82, Figs. 40–43. Botosaneanu & Gasith, 1971, *Israel J. Zool.*, 20:99, Fig. 3; Botosaneanu, 1973, *Fragm. ent.*, 9:61–80, Fig. 1; Botosaneanu & Giudicelli, 1981, *Proc. 3rd. int. Symp. Trichoptera*, (ed. Moretti), Series Entomologica, 20:21–29.
Hydroptila aïrensis Jacquemart, 1980, *Bull. Inst. r. Sci. nat. Belg. Ent.*, 52(13):1–5.

The following are figures obtained from about 100 specimens in a larger sample from Naḥal 'Arugot: wing expanse in the ♂ : 4.25–5.7 mm; in the ♀ : 5.3–6.2 mm; number of antennal segments, in the ♂ varying from 28 to 31, 30 and 31 being decidedly the most frequent numbers; in the ♀ this number is 24 in most specimens, seldom 23 or 25. Attached to the inner surface of the scent organ caps (Fig. 134), a single bunch of androconial setae (on the head, distally to the two scent-organs, another pair of such bunches).

Male genitalia: Segment IX, seen from side, very stout, proximally rather deeply excised on the dorsal side, very much less so on the ventral side, not produced anteriorly, with elongated and slender latero-posterior projections; sternite IX with a median excision of its posterior margin, this excision rounded, moderately deep, and bordered by very setose blunt projections. The elongated dorsal plate formed by segment X is mostly membranous, and thus difficult to describe; in any event, its apical margin is medially excised — thus bilobed; the most conspicuous feature of this plate is a strongly chitinized median ridge running almost to its apex, with an upturned black hook at its end. Beneath this dorsal plate there is another plate ("lower penis cover"?), subapically provided with a pair of spines. The long " ventral branches of segment X" are very characteristic: swollen basally, they are generally slender, not branched, and in their distal parts they are strongly elbowed (upturned), crossing each other, their apices being somewhat thickened and blackened. Inferior appendages, in ventral view, broad in their basal parts where they have a stronger lateral and a less strong median projection, then gradually narrowing towards the apices which are slightly capitate, blackened; in lateral view, the "shaft" of the gonopod is slender, but its main part is markedly widened (especially dorsad), the apex being directed

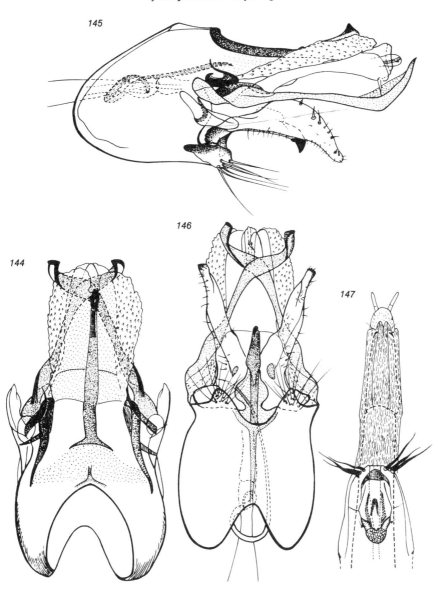

Figs. 144–147: *Hydroptila hirra* Mosely, 1948
144–146. male genitalia
(144. dorsal view; 145. lateral view; 146. ventral view);
147. female genitalia, ventral view.
(144–146: from Botosaneanu, 1973;
147: from Botosaneanu & Gasith, 1971)

obliquely downwards and with a small black "tooth". There is, between the gonopods, a very characteristic, long, strongly chitinized, dark median appendage, curving down like a hook at its apex.

Female genitalia: Terminally on the ventral side of abdominal segment VIII, there are two groups each of three strong setae, these two groups largely distant medially. The glandular organ within sternite VIII is very well sclerotized, its short lateral branches somewhat swollen, its central branch resembling a rounded, transversally developed, reticulated shield.

Remarks: Though belonging to the *occulta* group, *H. hirra* is a very peculiar and specialized species. Many characters of the genitalia (particularly, in the male, the sclerotized keel on the plate of segment X, the curiously shaped "ventral branches of segment X", the median appendage between the gonopods; and in the female, the shape of the glandular organ and of the row of apical setae of sternite VIII) will serve to distinguish it rapidly from any other Levantine species. Despite its extremely large distribution area, no interesting geographical variability could be observed. Only a brief mention should be made here of the fact that in a publication by Marlier & Marlier (1982) not less than 4 new species of *Hydroptila* were described from the island of La Réunion; according to the authors, these species "montrent des liens étroits entre elles et avec *H. hirra* Mosely ...". This seems to be an extremely interesting discovery, which will possibly lead to important biogeographical conclusions when this complex of species becomes better known. Mr F.M. Gibon (Bamako) informs me that *H. brincki* Jacquemart, 1963, described from Zimbabwe (Natal), seems to be practically identical with *H. hirra*; but Jacquemart's description and illustration are so unsatisfactory that no conclusion is, for the time being, possible. According to Malicky (1982/1983; 1986a), *H. hirra* is a junior synonym of *H. cruciata* Ulmer, 1912, described from Lake Nyasa, Tanzania; study of genitalia of the specimen thought to be Ulmer's type was not possible, but comparison with his drawings renders this idea plausible. For the time being, with the situation still unclear, I prefer to maintain *H. hirra*.

Distribution: First described from the mountains of the Yemen People's Republic, *H. hirra* was subsequently determined from the Sinai Peninsula, the Dead Sea Depression, one locality in Galilee, and also from the Aïr Mountains (Republic of Niger), from several countries of Western Africa, and from one of the Cape Verde Islands, São Antao. This is, indeed, a huge distribution area, including parts of the Oriental and of the Palaearctic regions, but especially of the Ethiopian region.

In the Levantine Province the species was found at Wadi Isla in the Sinai Peninsula at 700 m a.s.l. (22), in several localities situated at 130 to 350 m below sea level along the western shores of the Dead Sea e.g., from N to S — 'Ein Nueima (or 'Ein Duyuk), 'Ein Turabe, Nahal 'Arugot, 'En Mishmar (all 13), and also in the Galilee, as far north as Wadi (or Nahal) 'Ammud northwest of Lake Kinneret, at 100 m a.s.l.(7). It is the only species belonging to the "eremial complex" of Trichoptera, which reaches such a northern point, and this seems to be the northernmost locality in the distribution area of *H. hirra*.

Ecological Notes: In the Levant, this expansive and tolerant species mainly inhabits desert springs and spring-fed streamlets or larger streams; Naḥal 'Ammud, in the Galilee, is a brooklet with a rich growth of filamentous algae. Possibly the most opportunistic of all the species in the small "eremial complex", it sometimes develops very large populations and, though frequently accompanied by *H. adana*, is always more common than the latter species. Adults were caught in the Levant especially between September and December, but this does not seem relevant, because large numbers were also caught in April, in some localities of the Dead Sea Depression. Eco-ethological observations on the species were made in the spring complex of 'Ein Turabe and in a stream in the Aïr massif (Botosaneanu & Giudicelli, 1981).

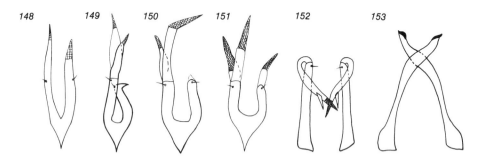

Figs. 148–153: Diagrammatic representation of
"ventral branches of segment X", male genitalia of
148. *Hydroptila mendli levanti*; 149. *H. palaestinae*; 150. *H. libanica*;
151. *H. fonsorontina*; 152. *H. adana*; 153. *H. hirra*

Hydroptila aegyptia Ulmer, 1963
Figs. 154–160

Hydroptila aegyptia Ulmer, 1963, *Arch. Hydrobiol.*, 59(2):267–268.
Hydroptila pulchricornis F.J. Pictet, 1834. Botosaneanu & Gasith, 1971, *Israel J. Zool.*, 20:99.
Hydroptila kurnas Malicky, 1974, *Ann. Mus. Goulandris*, 2:109–110.

Wing expanse, ♂: 4.4–5.5 mm (the unique available ♀ specimen being a metamorphotype, its wing expanse could not be measured). Antennae, in the ♂, with 28–31 segments; there is a geographical, or interpopulational, variability in this respect: in the Lebanese specimens, 29 is clearly the most frequently recorded figure; in specimens from 'En Te'o (Galilee) distinctly more specimens have 28, or 30 segments, than 29 segments; and the single specimen from 'En Tut (3) has 31 segments. In the single available female, the (extremely short) antennae have 21 segments.

Male genitalia: Abdominal segment VIII is modified: the sternite has posteriorly a very deep and wide triangular excision, and on its two halves there are dense, oblique

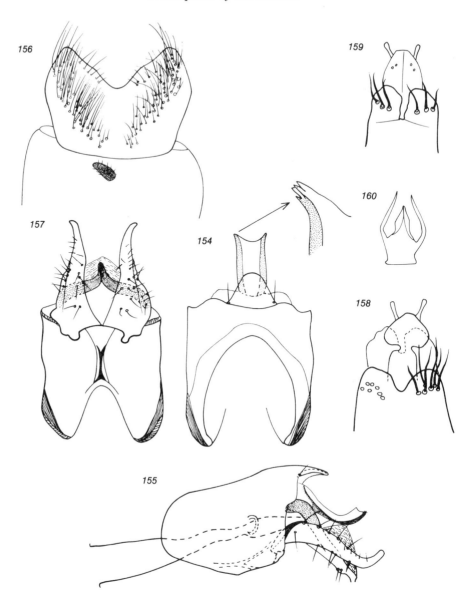

Figs. 154–160: *Hydroptila aegyptia* Ulmer, 1963
154–157. male genitalia
(154. segments IX & X, dorsal view, arrow pointing to
strongly magnified distal angle of "dorsal plate"; 155. lateral view;
156. segments VII & VIII, ventral view; 157. ventral view)
158–160. female genitalia
(158. dorsal view; 159. ventral view; 160. trident-like structure)

rows of long, densely ciliated setae. Segment IX, from the side, obtuse, its anterior end particularly obtuse, without true latero-posterior projections; tergite very much more deeply and broadly excised than sternite; the tergite has in the middle of its posterior border an important, bluntly triangular projection, perfectly delimited basally where it has a lateral seta on each side; the sternite also has a median projection, this being broad but short, angles rounded. From beneath the projection of tergite IX, a dorsal plate belonging to segment X arises: this is remarkably narrow, but moderately elongate, borders almost parallel and more strongly chitinized (darker) than the rest, apically emarginate, distal angles pointed (and actually represented by a cluster of minute points). The ventral part of segment X, directed obliquely down and backwards, has a very characteristic shape: its well chitinized parts are shaped like a "frame", its two lateral component parts are slightly sinuous, coalescent on the median line to form a common "head" terminating in a sharp "beak" directed downwards. Inferior appendages very large, of rather simple shape: in lateral view, strongly sinuous and rather slender, apical part obliquely upturned; in ventral view, slightly divergent, very broad in their basal halves, then strongly tapering to the divergent apices, lateral edges somewhat irregular. Phallus with a long constricted median part between two swollen parts, and with extremely short, coiled paramere.

The female of this species was never described. Abdominal segment VIII with sternite much longer than the tergite, both of them bilobed posteriorly, the lobes with long setae (but the excision separating the lobes of the sternite deep and very narrow, whereas that of the tergite is broader, rounded, less deep). Trident-like internal structure with extremely short, basally truncated stalk.

Remarks: *H. aegyptia* belongs to the small *pulchricornis*-group, and was sometimes confused with *H. pulchricornis* Pictet, 1834. It is possibly the only representative of the group in the Levant, and will be easily recognized by many peculiarities of the male and female genitalia. Apparently there is no interesting variability in the genitalia, but that in the number of antennal segments in the male was already mentioned.

Distribution and Ecological Notes: The species was described (Ulmer, 1963) from Egypt ("Maadi, Nilufer"); it was subsequently mentioned from Lake Kurnas, in Crete, from Lake Trasimeno, in Italy and from Tunisia. There is a possibility that it was mentioned, as "*pulchricornis*" from a small number of other localities, too. In the Levant, it was caught in two Lebanese localities: a small coastal river, Nahr ed Damour, 500 m downstream from the junction with Nahr el Hammam, and the Litani at Jib Jennine. In Israel, the species was caught exclusively in the Galilee, viz. in the 'En Tut spring, at 150 m a.s.l., near the foot of Mount Carmel (3), and in the spring-pool of 'En Te'o at 70 m a.s.l. (7).

The ecological spectrum of *H. aegyptia* is amazingly broad, indeed, because the species inhabits springs ('En Tut), large limnocrenic, spring-fed pools ('En Te'o), larger streams or small rivers belonging to the Metarhithral or to the Hyporhithral (Nahr ed Damour in its lower course, at 0–40 m a.s.l., where it is some 6 m broad; the Litani at 800 m a.s.l., where it is some 15–20 m broad), the largest rivers (the Nile in its lower course), and also shallow lakes (Trasimeno, Kurnas). Strangely enough, the distribution seems to be quite sporadic, possibly owing to biotic factors. Adults were caught, in the Levant, from March to September; the species is strongly attracted by artificial light. The almost complete absence of females in the samples (even in rather rich ones) is peculiar: possibly the only presently known female specimen is a "metamorphotype" caught in the Litani.

Hydroptila vectis Curtis, 1834

Figs. 161–168

Hydroptila vectis Curtis, 1834, *Phil. Mag.*, 4:217. Neboiss, 1963, *Beitr. Ent.*, 13(5/6):626.
Hydroptila maclachlani Klapálek, 1891, *Sber. K. böhm. Ges. Wiss.* (1890):177–181, 186, 190, Pl. 7, figs. 1–4, Pl. 8, figs. 1–8.
Hydroptila vectis var. *corsicanum* Mosely, 1930, *Eos, Madr.*, 6(2):176.

Wing expanse highly variable individually and between different populations: ♂ 4.9–6.8 mm, ♀ 4.9–8.9 mm; it is a large *Hydroptila*, and ♀♀ with a wing expanse of 8.9 to 9.5 mm, caught at Mas'ada in the Golan, are very large specimens for hydroptilids. Antennae, in the ♂, with 29–33 segments (interpopulational variability), in the ♀ with 24–27 segments (the antennae of the very large specimens are particularly short relative to the body length). On abdominal sternite VII in the male, an appendage which is short, hardly reaching the sternite's hind margin, or even much shorter.

Male genitalia: Segment VIII not excised. Segment IX very short, almost circular in lateral view, anterior excisions broad and not very deep; there is a medio-dorsal, posterior, semi-oval excision containing the stalk of the dorsal plate of segment X; the central parts of the sternite are only slightly chitinized (perhaps even membranous); characteristically, there are, on each side, three differently shaped closely spaced latero-posterior projections: the most dorsal one is the smallest, almost conical with irregular tip; that in the middle is the largest, a broadly rounded setose plate; the most ventral projection is long and digitiform, with strong setae basally. Dorsal plate of segment X only basally sclerotized, with narrower stalk and broad, rounded apical part. From beneath this plate, a pair of widely distant, symmetrical, long, black spines arise, which are almost horizontal and turned upwards and laterad in their apical part (these "elephant's tusks" are perhaps the most conspicuous part of the male genitalia). Inferior appendages consisting of a dorsal and a ventral part, which are articulated, forming a clasping complex; dorsal part with a relatively slender stalk with a baso-ventral spiny projection, and with a stronger distal part, downturned, and clearly capitate in dorso-ventral view, tip directed laterad; ventral part characteristically bifid, lateral branch longer and more slender than median branch. Phallus of complex appearance, with a long, coiled paramere with tip directed dextrally, and with inflated apical part giving rise proximally to a well developed hook directed sinistrally.

84

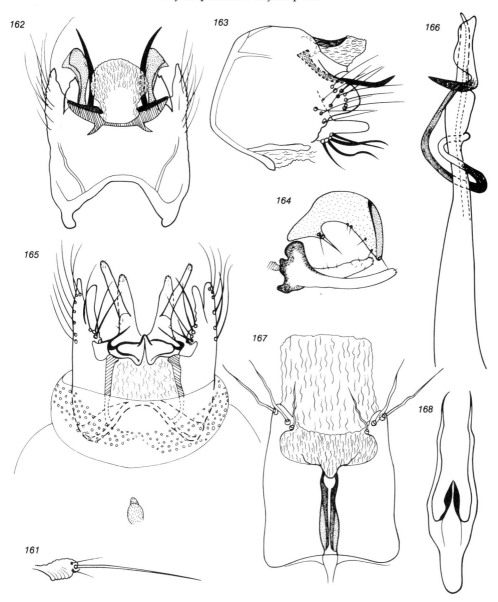

Figs. 161–168: *Hydroptila vectis* Curtis, 1834

161. setose tubercle on abdominal sternite V of male

162–166. male genitalia

(162. dorsal view; 163. segments IX & X, lateral view;

164. clasping complex formed by gonopod, lateral view; 165. ventral view;

166. phallic apparatus, ventral view)

167–168. female genitalia

(167. segment VIII, ventral view; 168. trident-like structure)

85

Female genitalia: Sternite VIII unmistakably characterized by a strong, dark median sclerotization, readily visible also in unprepared specimens. Trident-like internal structure with extremely elongated, slender lateral arms.

Remarks: This is a very peculiar species, having only one definite relative: *H. viganoi* Bots. These two species may be distinguished by a long series of evident features in the male genitalia.

The variability in size and in the number of antennal segments, has already been mentioned. Some variability was observed in the relative length of the two branches of the inferior part of the gonopod. This lead Mosely (1930) to describe a var. *corsicanum* with a very short median branch; however, this variable feature seems to be present in all the circummediterranean populations. More interesting seems to be the fact that the Levantine specimens are characterized by a short, or extremely short, appendage on sternite VII of the male: in many, or most, of the specimens from other zones, this appendage – as illustrated in several publications – is much longer, even reaching the posterior limit of segment VIII.

Distribution: *H. vectis* has a very large distribution area in the western Palaearctic, being known from almost the whole of Europe, northern Africa, the Levant, and also Pakistan. In the Levant, the species is abundantly distributed in Lebanon, where it was caught in many water courses belonging to the Orontes basin (Labwé, Jammouné, Chlifa, Baalbek), at 1,000–1,400 m a.s.l., and also in several water courses in the coastal basins of Nahr ed Damour, and especially of Nahr el Aouali, at 40–1,100 m a.s.l. It is present in Upper Galilee, where it was caught at 'En Qedesh at 460 m a.s.l. (1); in the Golan Heights (18), where it is known from Mas'ada; and in the Central Negev (17) where it was repeatedly caught at 475 m a.s.l. at 'En 'Avedat, one of the southernmost localities in the species' distribution area. More localities will undoubtedly be added, especially in the Galilee and the Judean Hills, but the occurrence of *H. vectis* in the arid zones south of the northern limit of the Dead Sea Depression must be an exceptional phenomenon (it has never been found in the Dead Sea area, despite careful collecting).

Ecological Notes: In Lebanon, the species appears to be a eurybiont, found from mountain springs and spring-brooks to rather large and sluggish watercourses at low altitudes (Crenal to Metarhithral); it is, nevertheless, also true that the most thriving populations inhabit psychrostenothermic, fast flowing springs and spring-fed streamlets and streams (for example: the spring-brook Aazibi in the Nahr el Aouali basin). In the Golan and Galilee the species is clearly a stenobiont restricted to the colder water of springs and brooks; its absence from localities apparently very propitious for its development could be the result of interspecific concurrence. The presence of a flourishing population in a desert spring with warm water (21–23°C) at 'En 'Avedat, in the Negev, is possibly exceptional: here the species has clearly madicolous habits; in fact such habits were apparently observed at some Lebanese localities, too. On the wing at least from March to October, it was also caught in January at 'En 'Avedat.

Hydroptila viganoi Botosaneanu, 1974

Figs. 169–174

Hydroptila viganoi Botosaneanu, 1974b, *Israel J. Ent.*, 9:160–164. Type Locality: spring at Wallaja, near 'Amminadav, Judean Hills, Israel. Type deposited in the Zoological Museum, Tel Aviv University.

Description of the ♂ (♀ presently unknown). Wing expanse ca. 6.4 mm. Antennae with 36 segments. Ventro-median appendage of segment VII pointed, reaching beyond the posterior limit of this segment. Sternite VIII with numerous long setae, posteriorly with deep and very broad triangular excision. Segment IX with a much broader anterior excision on the dorsal side than on the ventral side, these two excisions more or less equally deep; in lateral view, the segment is high distally, much less so proximally, proximal margin extremely oblique, dorsal margin almost horizontal, ventral margin sinuous, proximal end obtuse – ogival; there is, on each side, a pair of well developed but not very long latero-posterior projections with quite different shapes: the ventral projections, practically symmetrical, are ogival and very broad, furnished with long setae, whereas the dorsal projections (located immediately above the ventral ones) are asymmetrical, that on the left side being more developed and apically bidentate. The dorsal plate of segment X has a complex structure, not easy to understand: its apical part is a round, membranous lobe, there are also lateral membranous zones, and the central part of the plate is mainly well chitinized, but again with a membranous median zone. Below this plate, there arises a single very long, strong, black spine, situated on the right side; this spine is sinuous, only slightly directed laterad at its apex, and it is homologous to the pair of "elephant's tusks" found in *H. vectis*. Inferior appendages two-branched, dorsal and ventral branch connected (internally!) by a sclerotized ridge describing a perfect curve; dorsal branches capitate in lateral view, spatulate in ventral view, slightly divergent; ventral branches smaller, of very simple shape, slightly divergent, completely coalescent at base where they are furnished with several long setae, apices simply rounded. Phallus asymmetrically situated on the left side, anteapically with a rather short but strong, sinuous spine terminating in a kind of hook. Internally in the abdomen (perhaps corresponding to segment VII?) a large, globose, sclerotized formation (Fig. 174): the significance of this strange formation remains unknown.

Remarks: A very peculiar species, *H. viganoi* is related only to *H. vectis* Curt., these two species forming a small group and being readily distinguished from one another by a long series of good characters of the male genitalia.

Distribution and Ecological Notes: Up to the present only the type of this species is known. It was collected (31.III.1971) from a spring modified by man, at Wallaja, not far from the village of 'Amminadav, at 650 m a.s.l. in the Judean Hills (11); the specimen was caught on a vertical rocky surface constantly moistened by a thin film of spring water: it is not impossible that the species will be found to be a true madicolous element.

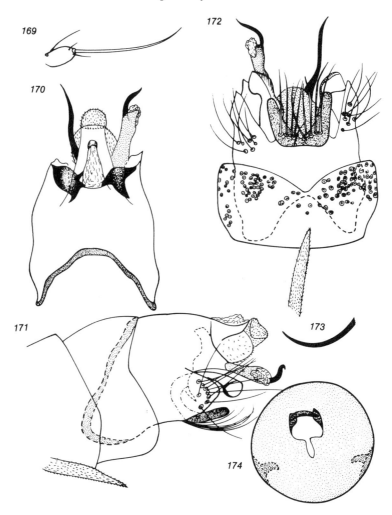

Figs. 169–174: *Hydroptila viganoi* Botosaneanu, 1974, male
169. setose tubercle on abdominal sternite V;
170–173. genitalia
(170. dorsal view, dorsal plate perhaps not quite correctly represented,
gonopods not represented; 171. lateral view; 172. ventral view;
173. the unique long black spine, lateral view);
174. intraabdominal sclerotized globose formation.
(From Botosaneanu, 1974b)

Genus OXYETHIRA Eaton, 1873

Trans. ent. Soc. London, 1873:130, 132, 143

Type Species: *Hydroptila costalis* Curtis, 1834, sensu Eaton, 1873, or *Oxyethira flavicornis* (Pictet, 1834). But "the true identity of the type-species is still in dispute", and "there is still not an unequivocal solution to the problem" (see discussion in Neboiss, 1963:594–595 and in Marshall, 1979:203–204).

Diagnosis: Hydroptilids best characterized by the combination of the following characters: 3 ocelli present; spur formula 0, 3, 4; wings very narrow and strongly pointed at apex (i.e., with long apical lobes). There are extremely long fringes on the costal margin in both wings, and on the anal margin in the hindwings. Venation in both wings indistinct and strongly reduced; in the forewing, SR and M both have three branches, each branch forming one apical fork; in the hindwing SR is not branched, but M is apically branched, forming what looks like a fork (there are important variations in the wing venation at the specific — or subgeneric — level, and Fig. 175 cannot be considered as representative of the whole genus). Laterally on sternite V a pair of minute setose bulbs corresponding to internal glands. In the male, a medio-distal pointed process on abdominal segment VII; in the female such a process present on segment VI. Male genitalia symmetrical without strongly protruding parts (phallic complex excepted), difficult to interpret without careful preparation and observation; they are diversely shaped in the various subgenera and species-groups, and the description which follows, in which the terminology in Kelley (1984, 1985) is followed, mainly applies to the species-group to which the Levantine species belong. Segment VIII with tergite and sternite generally fused, apical margins dorsally and especially ventrally excised, laterally with blunt processes ("lateral processes") armed with conspicuous black, strong spines. Segment IX completely or almost completely withdrawn in segment VIII, often with deep proximal excision (lateral view!) dividing it into a shorter dorsal and a longer ventral part; the segment has "postero-lateral processes". In the distal part of sternite IX, always a median sclerotized lobe of various shapes, often moderately protruding, interpreted as "fused inferior appendages"; this median lobe frequently has on both its sides a pair of "postero-lateral lobes of venter IX", and these 3 lobes form characteristic complexes; a pair of membranous, transparent "setal lobes", apically with several short setae, are seen to be associated with the interior (mesal) surface of the "fused inferior appendages"; on a more dorsal level there is another pair of membranous, transparent formations, the "bilobed processes": they are generally larger, digitiform, inflated, tipped by one seta, and associated with the ventral face of the "subgenital plate". This plate is a heavily chitinized, dark formation, easily discernible just beneath the phallic complex; in most species, a pair of "subgenital processes" protrude from the apices of the two halves of the plate, being directed ventrad and seemingly forming a clasping system with the "fused gonopods" (these processes are sometimes interpreted as the dorsal part of the gonopods). There is no general agreement as to what the contribution of segment X is to the structure of the male genitalia. Female genitalia short (not forming an oviscapt); segment VII not modified; segment VIII with the short, arched tergite bearing anteriorly a pair of short and robust apodemes, and having in the centre of its apical margin a button-like projection with minute spines; segment IX strongly reduced; segment X a dome-like plate with the two cerci; several internal sclerites, offering the poor distinctive specific characters, are found in segment VII (spermathecal sclerite) or in segment VIII (the "horizontal lamella", and, posterior to it, a pair of large "C-shaped sclerites" on the floor of the oviduct).

General Remarks: About 110 species of this large genus with world-wide distribution are presently known (Kelley, 1984); they are classified in 10 subgenera, some of them being further subdivided into species-groups (some species are still "incertae sedis"). The 3 known Levantine species belong to subgenus *Oxyethira* and to the *falcata*-group which includes about 14 species, most of them western Palaearctic, but with two Nearctic species and two species in Madeira or in the Canary Islands. The discovery of some additional species, especially in the northernmost parts of the Levantine Province, cannot be excluded. Generally speaking, the species are associated with a vast spectrum of lenitic and lotic habitats, but all 3 Levantine species are typical running-water inhabitants (Crenal, Rhithral), with a more or less marked preference for cool water without a strong current and with a rich vegetal growth including filamentous algae; altitudinal range more or less broad; the species are never found in the desert zones. The larvae build characteristic minute cases: bottle-shaped, laterally compressed, semi-transparent, constructed of secretion only, posterior end broad with slit-like opening, anterior end narrower and thickened around the opening; the pupal case is the sealed larval case horizontally fixed to the substrate by four filaments ending in adhesive discs.

Key to the Species of Oxyethira in the Levant

♂♂

1.   Prominent projections on the sides of the dorsal excision of segment VIII; segment IX (lateral view) not divided into upper and lower part.   **O. delcourti** Jacquemart

–   No projections on the sides of dorsal excision of segment VIII; segment IX (lateral view) deeply divided into upper and lower part   2

2.   In the ventro-distal part of segment IX, a distinctive complex of 3 lobes, the median one longer, distally blackened; sclerotized "subgenital plate" with its medio-apical angles bluntly bidentate.   **O. falcata** Morton

–   Median lobe in ventro-distal part of segment IX obscured, being folded against the dorsal face of the horseshoe-shaped complex formed by the lateral lobes; "subgenital plate" (lateral view) simply rounded   **O. assia** Botosaneanu & Moubayed

♀♀

(Key to be used with caution: possibly not always reliable)

1.   "C-shaped sclerites" (segment VIII) with distal part as broad as proximal part, partly obscured by a pair of large, triangular plates developed at a lower level

                 O. delcourti Jacquemart

- "C-shaped sclerites" with distal part more slender than proximal part, not obscured by other formations   2
2. "C-shaped sclerites" not leaning on the "horizontal lamella" which is rather widely separated from the spermathecal sclerite   **O. assia** Botosaneanu & Moubayed
- "C-shaped sclerites" widely separated from the "horizontal lamella" which leans on the spermathecal sclerite.   **O. falcata** Morton

Oxyethira (Oxyethira) delcourti Jacquemart, 1973
Figs. 175–181

Oxyethira delcourti Jacquemart, 1973, *Bull. Inst. r. Sci. nat. Belg. Ent.*, 49(4): 5–9. Kelley, 1985, *Trans. Am. ent. Soc.*, 111: 240–242.

Wing expanse: 5.1–5.6 mm in the ♂, 5.6–6.5 mm in the ♀ (but a ♀ with a wing expanse of only 5.2 mm was also observed). Antennae, in the ♂ with 32–37 segments, in the ♀ with 25–27 segments. Wings: Fig. 175.

Male genitalia: Segment VIII much longer dorsally than ventrally, posterior margin of tergite with shallow emargination on both sides of which there are important "dorso-lateral processes", conspicuous especially in lateral view; "lateral processes" not strongly protruding, each bearing 2 strong, black spines, which are curved and divergent. Segment IX stout, shorter than segment VIII, with "shoulders" (ventral view), in lateral view without distinctly separated tergite and sternite, i.e. without any excision of proximal margin; "postero-lateral processes" of segment IX not strongly protruding. The most striking feature is the strongly protruding medio-distal lobe of the sternite ("fused inferior appendages"): in lateral view this looks like a very strong, darkish horn directed backwards and, before its tip, strongly upwards; in ventral view its lateral, strongly chitinized parts converge towards an elongate, blunt "head"; there is nothing on the sides of this formation which could be interpreted as "postero-lateral lobes". "Setal lobes" relatively large, easily observed, somewhat quadrangular, distal margin with at least 4 setae. "Bilobed processes" distinctly larger than "setal lobes", with one strong seta. "Subgenital processes": from a large, strongly chitinized, dark, complex plate, a pair of processes arise, which are a second striking feature of these genitalia. These processes are slender, in lateral view resembling elongate hooks curved ventrad to form apparently a clasping system with the "fused gonopods", in ventral view slightly converging, tips pointed mediad. Phallic complex with apical part capitate, dorsal angle bluntly protruding upwards, without any serrate keel.

Female genitalia built as in other *Oxyethira*. The "C-shaped sclerites of floor of oviduct" are equally broad in their proximal and distal parts; they lean on the "horizontal lamella"; the most distinctive feature is the presence in the ventral part of segment VIII of two very large, triangular, dark plates, situated on a lower level than the "C-shaped sclerites", and more or less obscuring them.

Remarks: *O. delcourti* is a member of the *falcata*-group, and is very closely related to *O. mithi* Malicky, a species known from Crete.

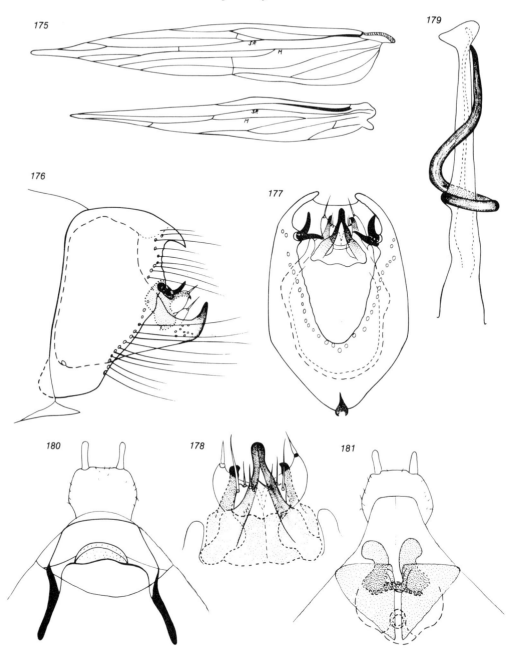

Figs. 175–181: *Oxyethira delcourti* Jacquemart, 1973
175. wings; 176–179. male genitalia (176. lateral view; 177. ventral view;
178. more strongly magnified view of apical parts of segment IX,
ventral view; 179. phallic apparatus, lateral view);
180–181. female genitalia (180. dorsal view; 181. ventral view)

Distribution: The species was described from Rhodes, and was subsequently also recorded from Kithira, Crete, Cyprus, and Asia Minor; the record from Crete is probably erroneous, and concerns the related *O. mithi*; *O. delcourti* will probably be found on other islands in the Aegean Sea. In the Levant it is restricted to Lebanon, and apparently (?) to the small coastal basins; it was caught here in the main course of Nahr ed Damour, at 260 m and at 40 m a.s.l., in two tributaries of this small river, at 260 m and at 45 m a.s.l., in the main course of Nahr el Aouali, from 380 to 50 m a.s.l., and in two tributaries of this watercourse, at 990 m and at 36 m a.s.l. There is also a questionable record from the lower course of the Litani, below 200 m a.s.l. Ecological Notes. Found in smaller or larger streams, this rhithrobiont possibly attains its ecological optimum in the main watercourses (Metarhithral), below 500 m a.s.l., inhabiting mainly calm zones with abundant growth of filamentous algae; intermittent streams are also inhabited; the presence of *O. delcourti* in Nabaa Aazibi (a spring and spring brook at rather high altitude in the basin of Nahr el Aouali) seems somewhat exceptional. Adults were caught from May to September, but the species is certainly also on the wing later in the year, ripe pupae being taken in November.

Oxyethira (Oxyethira) falcata Morton, 1893

Figs. 182–187

Oxyethira falcata Morton, 1893, *Trans. ent. Soc. London*, pp. 80–81. Kelley, 1985, *Trans. Am. ent. Soc.*, 111:242–245.
Oxyethira rhodani Schmid, 1947, *Mitt. schweiz. ent. Ges.*, 20(5):531–532.
Oxyethira bidentata Nybom, 1948, *Commentat. biol.*, 8(14):9–10 (renamed by Nybom in 1954 as *dentata*).

In a sample from 'En Tut, Israel, the wing expanse in the ♂♂ was 5.5–6.7 mm; in the ♀♀ 5.3–9.1 mm (exceeding 6 mm in most of the specimens). Male antennae with 36–40 segments, ♀ antennae with 25–30 segments; ♂ antennae conspicuously longer than those of the ♀.
Male genitalia: Segment VIII of about the same length dorsally and ventrally, posterior margin of tergite with deep excision on both sides of which the segment margin is slightly crenulate; "lateral processes" strongly protruding, each bearing 2 — but sometimes 3 — strong, black spines, which are almost straight. Segment IX slender, almost as long as segment VIII, with "shoulders" (ventral view); in lateral view there is a very deep excision of the proximal margin, i.e., there is a distinct separation of the dorsal and of the much longer ventral part, which are equally high (broad); "postero-lateral processes" of segment IX protruding. The most striking feature is the complex, in the ventro-distal part of segment IX, formed by the medio-distal lobe of the sternite ("fused inferior appendages") and the "postero-lateral lobes" located on both its sides; the medio-distal lobe is distinctly more protruding than the postero-lateral lobes, ogival with an apical nipple, distal parts blackened; the

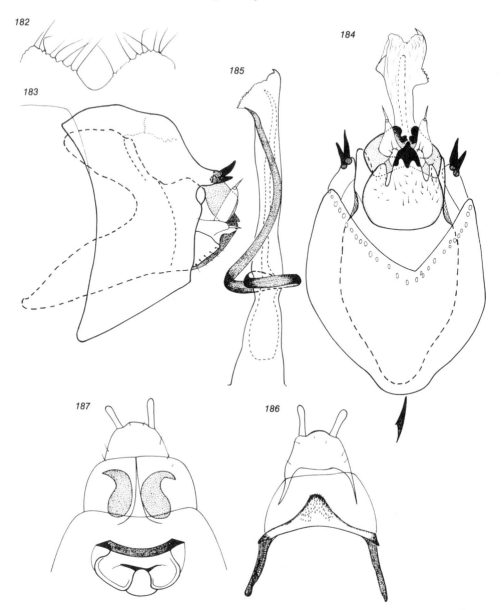

Figs. 182–187: *Oxyethira falcata* Morton, 1893
182–185. male genitalia
(182. posterior border of tergite VIII, dorsal view; 183. lateral view; 184. ventral view;
185. phallic apparatus, lateral view);
186–187. female genitalia
(186. dorsal view; 187. ventral view)

lobes on its sides are blunt, but not broadly rounded (somewhat triangular in lateral view). "Setal lobes" relatively small, irregular, with about 3 short setae. "Bilobed processes" distinctly larger than setal lobes, with one short spine. "Subgenital processes" represented by a large, strongly chitinized plate with a medio-apical excision on both sides of which the plate is greatly thickened, bluntly bilobed (these blackened lobes distinct in lateral view, too, and apparently forming with the "fused gonopods" a kind of clasping system). Phallic complex with apical part capitate, with two serrate keels and a small sharp beak at one of the apical angles.

Female genitalia built as in other *Oxyethira*. The "C-shaped sclerites" are perfectly distinct, their apical part much more slender than the proximal part; they are widely separated from the "horizontal lamella" which has sharp lateral angles and is highly adpressed to (coalescent with ?) the spermathecal sclerite.

Remarks: A member of the species-group bearing its name, *O. falcata* is not very closely related to certain other Levantine species. There are as yet no reliable published observations concerning interesting geographical variation, although there are statements about the existence of "eco-biogeographic forms" in this species.

Distribution: *O. falcata* has a vast Western Palaearctic distribution, being known from most of Europe, the Azores, North Africa, Asia Minor, Iran, but also from Pakistan. It was mentioned from Jordan, but without indication of localities. In Lebanon it was caught at Labwé (Orontes basin, 1,000 m a.s.l.), in a large tributary of the Litani (Ghozayel, 900–1,000 m a.s.l.), and in the middle reach of a small coastal river, Nahr el Aouali, at 380 m a.s.l. Several Israeli localities are in Upper Galilee (1), between 200 and 460 m a.s.l.: Tel Dan (Tel el Kadi), 'Ein Habis which is one of the springs of Nahal Keziv, and 'En Qedesh; moreover, it is known from the Golan Heights (18), being caught at Kafer Naffakh, at 700 m a.s.l.; and from 'En Tut, near the foot of Mt. Carmel, at 150 m a.s.l. (3).

Ecological Notes: A distinctly crenophile species — and not only in the Levant; most of the known Levantine localities are springs and spring brooks. However, the species does not avoid streams, including the Hyporhithral. It apparently needs cool water and a slight to moderate current. In the Levant it was found on the wing almost throughout the year.

Oxyethira (Oxyethira) assia Botosaneanu & Moubayed, 1985
Figs. 188–193

Oxyethira assia Botosaneanu & Moubayed, in: Moubayed & Botosaneanu, 1985, *Bull. zool. Mus. Amsterdam*, 10(11):65–67, Figs. 5–11. Type Locality: Labwé, Lebanon. Type deposited in the Zoological Museum, University of Amsterdam.

Wing expanse of ♂ holotype: 6 mm; of two ♀♀: 10.8 and 13 mm (a large *Oxyethira*!). Number of segments of the antennae unknown.

Male genitalia: Segment VIII of about the same length dorsally and ventrally, posterior margin of tergite with deep triangular excision on both sides of which the

95

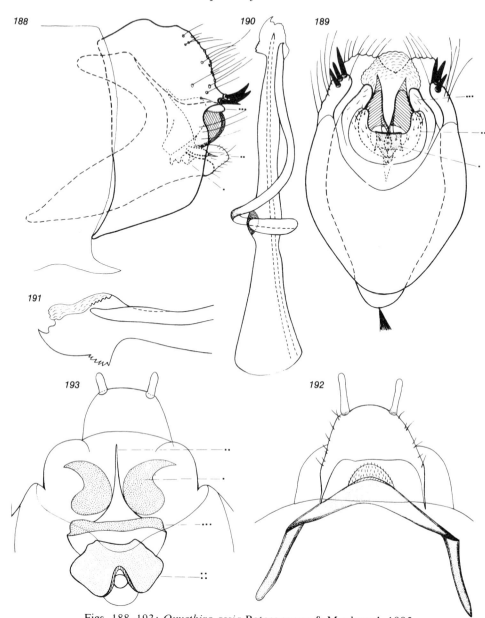

Figs. 188–193: *Oxyethira assia* Botosaneanu & Moubayed, 1985
188–191. male genitalia (188. lateral view; 189. ventral view;
190–191. phallic apparatus, and its apical parts more strongly magnified;
in Figs. 188–189, fused inferior appendages are marked by one dot, postero-lateral lobes
of venter IX by two dots, postero-lateral processes of segment IX by three dots);
192–193. female genitalia (dorsal view, ventral view; in Fig. 193, "C-shaped sclerites"
are marked by one dot, "V-shaped process" by two dots, "horizontal lamella"
by three dots, spermathecal sclerite by four dots).
(From Moubayed & Botosaneanu, 1985)

segment margin is slightly crenulate; "lateral processes" strongly protruding, each bearing 3 strong, black spines which are straight or slightly curved; this segment lacks a lateral suture. Segment IX slender, even longer than segment VIII, without "shoulders" (ventral view); in ventral view there is a very deep excision of the proximal margin, i.e., there is a distinct separation of the dorsal and of the much longer ventral part, the dorsal part being less high (broad) than the ventral part, but nevertheless moderately robust; "postero-lateral processes" of segment IX protruding. The medio-distal lobe of sternite IX ("fused inferior appendages") characteristically does not protrude at all, but, on the contrary, is folded on the dorsal side of the complex formed by the "postero-lateral lobes"; its shape is complex, and it can be observed only through the transparent "postero-lateral lobes", which form a well-developed, semicircular complex, the lobes proper being slightly convergent towards their broadly rounded apices (seen laterally, their outline is slightly irregular). "Setal lobes" very small, with one seta. "Bilobed processes" practically hidden behind the "postero-lateral lobes" of venter IX, and not represented in our drawings. "Subgenital processes" relatively simple: rounded in lateral view, sharply pointed at apex in ventral view. Phallic complex with apical part capitate, with two serrate keels and a small sharp beak at one of the apical angles.

Female genitalia built as in other *Oxyethira*. The "C-shaped processes" very distinct, their apical part much more slender than the proximal part; they are separated from the "horizontal lamella" which is rather widely separated from the spermathecal sclerite.

Remarks: Belonging to the *falcata*-group of species, *O. assia* is certainly the sister species of *O. boreella* Svensson & Tjeder, described (Svensson & Tjeder, 1975) from northern Sweden and presently known only from there. Only the following characters of *O. assia* allow its separation from *O. boreella*: size considerably larger; in the male — segment VIII without lateral suture; tergite IX (lateral view) stronger; segment IX in ventral view perhaps more slender; the lateral and ventral shape of the "postero-lateral lobes of sternite IX" which are broadly rounded at apices, not triangular; in the female — "C-shaped sclerites" with distal part more slender than proximal part; "horizontal lamella" not adpressed to (fused with) spermathecal sclerite.

Distribution and Ecological Notes: Presently known only from two localities in the upper basin of the Orontes, Lebanon: cool karstic springs with slight to moderate current, at Labwé (1,000 m a.s.l.) and at Jammouné (1,300–1,400 m a.s.l.). It is possibly a true crenobiont of (higher) mountains. Its coexistence with *O. falcata* at Labwé should be noted. Anyway, it must be a rare species.

Genus ORTHOTRICHIA Eaton, 1873

Trans. ent. Soc. London, 1873:130,132, 141

Type Species: *Hydroptila angustella* McLachlan, 1865, by original designation.

Diagnosis: Medium-sized hydroptilids; without ocelli; postoccipital warts on the head globose, located in small holes, not firmly attached to substrate, but also not transformed into scent organs. Antennae rather long for hydroptilids, not moniliform. Spur formula: 0, 3, 4. Forewings moderately narrowed, hindwings very narrow, both not very produced at apex, very setose and also with numerous erect setae, fringes on hindwings particularly long; venation reduced but not strongly (in forewing SR with 4 branches = apical forks 1 and 2, M with 3 branches = fork 3; in hindwing SR with 3 branches = fork 2, M with only 2 branches = no apical fork; this situation certainly not present in all *Orthotrichia*); in forewing of some species, a subcostal row of black androconial scales. On abdominal venter V sometimes a pair of setose warts corresponding to internal glands (both sexes). On abdominal sternites VI and VII of the male, and on sternite VI of the female, unpaired, short appendages (often broad, flattened), mostly concealed by setae. Male genitalia extremely asymmetrical in all their parts and in all species; the interpretation of the different parts of segments IX and X is difficult, and my descriptions will be rather pragmatic (for complete descriptions of male and female of *O. costalis*, see Nielsen, 1957:67–76, and Nielsen, 1980:100–105). Segment VIII generally not modified, symmetrical. Segment IX a stout, sclerotized capsule, deeply excised laterally; on the dorsal side, in its direct continuation, there is a "dorsal plate" certainly belonging to segment X, and furnished with various asymmetrical formations and appendages. On the distal border of sternite IX there are — in the *angustella*-group — diversely shaped asymmetrical appendages, the central one (an inconspicuous plate) being interpreted as "fused inferior appendages" (in *O. costalis* the gonopods are, instead, laterally placed, widely separated, elongate); the apico-lateral angles of segment IX produced in conspicuous appendages, sometimes unilaterally absent (note that the gonopods in *O. costalis* should not be confused with the apico-lateral appendages!). On levels intermediate between "dorsal plate" and venter IX, there is a "bilobed process", and from the complex internal skeleton of sclerotized ridges of the genital capsule, a long spiniform appendage arises. Phallus complex very long and slender, with a shorter or longer titillator. Female genitalia short, not forming a true ovipositor, of complex constitution; best developed of all abdominal segments is the unmodified segment VII; segment VIII a sclerotized ring with anterior apodemes, posteriorly with stronger setae, ventrally with often asymmetrical opening of the duct of a gland; segment IX with anterior apodemes on its ventral side often with various sclerites.

General Remarks: A very specialized, large genus present in almost all parts of the world (perhaps not in South America). In 1979, Marshall listed 40 species for the world fauna; but, that same year a publication by Alice Wells appeared, with a first report of the genus from Australia, with 25 species, almost all new! Several species-groups are recognized; two Levantine species, *O. moselyi* and *O. melitta*, belong to the important *angustella*-group, the third one, *O. costalis*, to the much smaller group bearing its name; both groups are "Old World"; the genital differences (♂ , ♀) between these two groups are considerable, and there are also non-genital differences (presence/absence in the male forewing of a subcostal row of black scales; presence/absence in both sexes of a pair of minute antero-lateral setose warts on venter V). In general terms, *Orthotrichia* comprises mainly inhabitants of lakes, ponds, and marshes, but a number of species are known (or known also) from the lenitic habitats of running water with abundant algal growth (probably never truly cold, fast-running, or situated at high altitudes). The young instars have highly characteristic cases made of silk alone, resembling cumin seeds and with typical dorsolateral longitudinal ridges (keels) on each side.

Key to the Species of Orthotrichia in the Levant

♂♂

1. Forewing without conspicuous black line (formed by androconial scales) along basal part of subcosta; inferior appendages well protruding, widely separated, arising laterally from sternite IX. **O. costalis** (Curtis)
–   Forewing with such a line; inferior appendages fused, forming at posterior margin of sternite IX an inconspicuous centrally located plate 2
2. Basally from left side of the "dorsal plate" a digitiform appendage with two setae, not crossing the plate; latero-apical projections of segment IX markedly asymmetrical. **O. moselyi** Tjeder
–   The appendage arising basally from left side of the "dorsal plate" is a strong horn obliquely crossing the plate; latero-apical projections of segment IX only slightly asymmetrical. **O. melitta** Malicky

♀♀

1. Stronger setae on segment VIII restricted to two latero-posterior sclerotized strips; centrally on ventral side of segment IX a pair of sclerites. **O. costalis** (Curtis)
–   A row of strong setae along most of posterior border of segment VIII; centrally on ventral side of segment IX, either no sclerites, or one large sclerite 2
2. Centrally on ventral side of segment IX no sclerite; there is some asymmetry on the ventral side of segment VIII. **O. moselyi** Tjeder
–   Centrally on ventral side of segment IX one large sclerite; ventral side of segment VIII practically symmetrical. **O. melitta** Malicky

Orthotrichia moselyi Tjeder, 1946

Figs. 194–203

Orthotrichia moselyi Tjeder, 1946b, *Opusc. Ent.*, 11:133–134. Gasith & Kugler, 1973, *Israel J. Ent.*, 8:57–58. Type Locality: "Palestine, Dagania A., Jordan Valley, 670 ft. below the sea". Type deposited in the British Museum (Natural History).

Wing expanse in the ♂♂ 4.1–7.1 mm (anterior wing in the holotype 2.75 mm, giving an expanse of ca. 8.5 mm); in the ♀♀ 5.7–6.8 mm; these figures possibly do not quite correctly reflect the real situation. Antennae of ♂ with 29–32 segments, those of ♀ with 27–28 segments. Head with conspicuous tufts of whitish setae. General tint of setae on the wings reddish-copper; long, dense fringes on both wings, anal fringes of hindwings particularly long, in forewings apical half of fringes with alternating patches of dark and pale setae; numerous long, erect, pale setae in longitudinal rows on forewing. In the ♂ (Fig. 199) a very conspicuous row of black androconial scales present along the basal half of the subcosta in the forewing, more easily seen in rubbed alcohol preserved specimens). Close to the anterior border of segment V a pair of minute setose lateral warts (corresponding to internal glands).

Male genitalia completely asymmetrical. Proximal excision of segment IX much larger (higher) on right side than on left side. Irregularly shaped, long, moderately narrow "dorsal plate" (segment X); from left to right it features in its apical part: a slightly irregular apex; a very pale, ovoidal, membranous(?) lobe; a well chitinized, rounded lobe whose margin is produced into a sclerotized horn-like process, not very long (perhaps twice the lobe diameter) and obliquely directed downwards; further basad a long digitiform appendage with 2 apical setae. In ventral view at the distal margin of segment IX, a complex and quite asymmetrical set of appendages, perhaps all belonging to this segment; the central position is occupied by a plate, apically excised, asymmetrically bilobed: it is interpreted as medially fused inferior appendages; to its left is a dark, apically hook-shaped appendage, having as annex a digitiform appendage; both apico-lateral angles are produced in highly asymmetrical appendages, with setae apically: on the right side is located the longest and strongest of all formations, slightly sinuous, basally accompanied by a strong conical projection (with a much smaller counterpart on the opposite side); on the left side is an asymmetrical pair of shorter, digitiform appendages starting from a common root. In lateral view, the lateral appendages and also the dark, hook-shaped one are best seen, the fused gonopods being very inconspicuous. At a level intermediate between the dorsal plate and venter IX, a pair of slender appendages (the "bilobed process"). Phallic complex extremely long and slender, with one sinuous, moderately long, not strongly coiled titillator (about 1/3 of the length of phallus).

Female genitalia (never before described): Segment VII not modified; it is the largest abdominal segment. Venter of segment VIII asymmetrical, owing (not exclusively!) to the apical part of the gland duct with its large opening being turned to the right; a row of long, curled setae near the segment's posterior border is only medio-dorsally and medio-ventrally interrupted. Venter IX characterized by the absence of any sclerite, and by the presence of a pair of very well-developed, bluntly conical and membranous lobes, with minute spinules apically; there is on the sides of segment IX a pair of moderately long, very slender (but broader basally) pale appendages.

Remarks: *O. moselyi* belongs to the *angustella*-group of species, like *O. melitta* (these two species are very closely related, probably sister species).

100

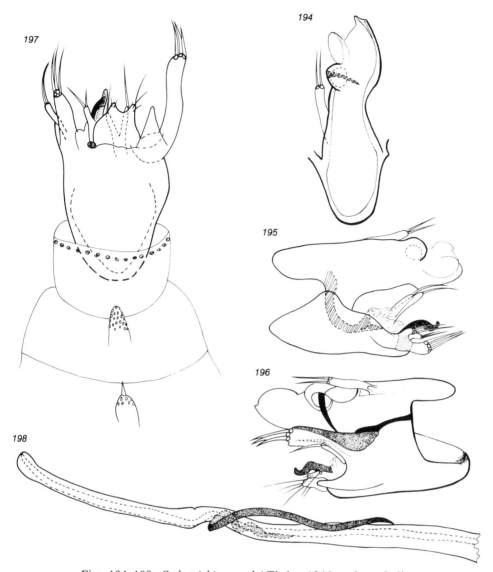

Figs. 194–198: *Orthotrichia moselyi* Tjeder, 1946, male genitalia
194. segments IXth & Xth, dorsal view;
195. lateral view, left side; 196. lateral view, right side; 197. ventral view;
198. distal part of phallic apparatus

Distribution: This species is possibly a Levantine endemic element. It was caught in Lebanon in one locality in the Orontes basin (downstream from the main Zarka spring, at Hermel, 650 m a.s.l.) and in one locality in the Litani basin (Ghozayel, tributary of Litani, at Anjar-Chamsine, 900–1,000 m a.s.l.). In Israel, most of the localities are around Lake Kinneret, at ca. 200 m below sea level (7); but it is also

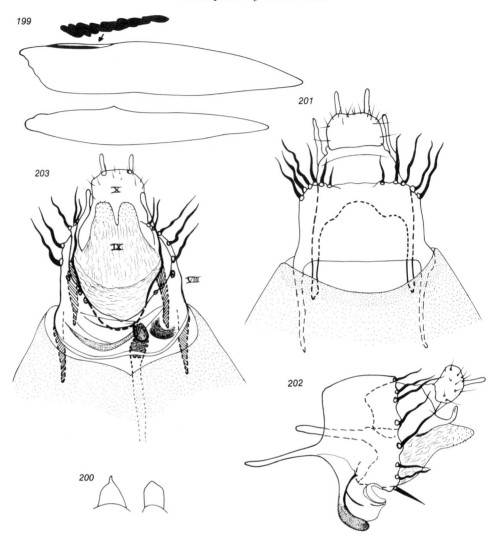

Figs. 199–203: *Orthotrichia moselyi* Tjeder, 1946
199. wings of male, without neuration
(above forewing: some androconial scales from the subcostal row);
200. median appendage on abdominal sternite VI, two different specimens;
201–203. female genitalia
(201. dorsal view; 202. lateral view; 203. ventral view)

known from 'En 'Amal (Kibbutz Nir Dawid), at ca. 100 m a.s.l., in the Valley of Yizre'el (5). Surprisingly enough, specimens were also collected at Wadi Isla, in the Sinai, at 700 m a.s.l. (22). I strongly suspect that it will also be discovered in suitable localities in Syria (e.g., Lake Homs), and in Jordan.

Ecological Notes: *O.moselyi* is mainly a limnobiont, the largest known population(s) being in Lake Kinneret. According to Gasith & Kugler (1973), large numbers of adults were caught here in light traps from mid-March to October, with two (May and September–October) or more peaks each year, pointing to at least two generations annually; adults were absent from January to March. However, the species is not, strictly speaking, a limnobiont, as shown by the fact that it was found in lotic habitats in all the other known localities: a spring with slow flowing water at 'En 'Amal; meanders with slight current, of the Orontes, 10 m broad at Hermel; a rather large river (10–15 m broad) like Ghozayel, and probably also karst springs in its basin; and desert running water courses in the Sinai, at Wadi Isla. At all these localities the populations are certainly much smaller than in Lake Kinneret. It should be noted that at one of them ('En 'Amal) adults were caught as early as February.

Orthotrichia melitta Malicky, 1976
Figs. 204–210

Orthotrichia melitta Malicky, 1975 (1976), *Z. ArbGem. öst. Ent.*, 27(3/4):93, Taf. 2. Kumanski, 1985, *Fauna na Balgariia*, 15 (Trich., Annulipalpia), pp. 148–150.

Wing expanse in the two specimens from Lebanon: ♂ 4.7 mm, ♀ 4.9 mm (length of forewing mentioned in publications for specimens from Lesbos and Bulgaria: 2.5–3 mm, thus an expanse of ca. 5.3–6.4 mm). Number of antennal segments, in Bulgarian specimens: ♂ ca. 32, ♀ ca. 27. Subcostal row of black androconial scales well developed. Abdominal venter V in both sexes with antero-lateral pair of warts, with one very long and some very short setae.

Male genitalia completely asymmetrical. Proximal excision of segment IX much larger (higher) on right side than on left side. Dorsal plate (segment X) very irregularly shaped, elongate, left apical angle produced into a strong but short upturned sclerotized spine; on the right side there is a strongly chitinized rounded lobe whose margin is produced into an inconspicuous spine; this lobe is immediately followed by a second one, less dark; further basad there arises from the plate's margin a long sclerotized appendage with acute apex and without setae, directed backwards and obliquely crossing the plate: this very characteristic appendage is obviously homologous with the long digitiform appendage with 2 apical setae observed in *O. moselyi*. In ventral view at the distal margin of segment IX, the central position is occupied by two formations with tips almost touching each other (Fig. 208): (a) on the right side is located a massive sclerotized formation, not bilobed, with narrowed tip turned to the left; this (interpreted as "fused inferior appendages") is separated by a triangular projection from (b) a dark, sclerotized complex made by a slender, strongly sinuous appendage (more or less hook-shaped in lateral view), with a very irregularly shaped annex with several setae. The apico-lateral angles of segment IX are produced in long and broad appendages which are only slightly asymmetrical, not

103

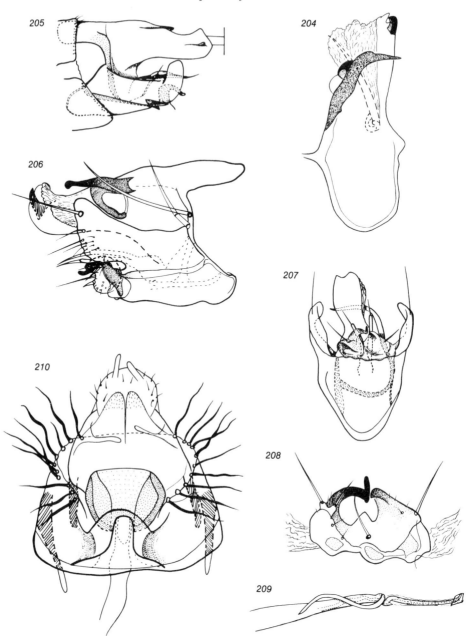

Figs. 204–210: *Orthotrichia melitta* Malicky, 1976
204–209. male genitalia (204. segments IX & X, dorsal view;
205. lateral view, left side; 206. lateral view, right side; 207. ventral view;
208. appendages on central part of ventro-posterior margin of
segment IX, strongly magnified; 209. phallic apparatus);
210. female genitalia, ventral view. (205, 207, 209: from Kumanski, 1985)

split, without annex formations, both distally bent upwards and mediad, apical part with many setae, medially with large membranous zones. Between dorsal plate and venter IX, a "bilobed process" whose lobes are extremely long and slender; Fig. 207 also features a long, movable, spiniform appendage arising from the complex skeleton of sclerotized internal "ridges" of the genital capsule (I am unable to interpret it; it probably occurs in all species of the genus, but I did not illustrate it in *O. moselyi*). Phallic complex as in *O. moselyi*.

Female genitalia: Venter of segment VIII practically symmetrical owing to the apical part of the gland duct with its (not enlarged) opening having a median position; a row of long, curled setae near the segment's posterior border is dorsally and ventrally interrupted. Venter IX with a very characteristic large sclerite (polygonal-elliptical); posterior to it, a pair of large, bluntly conical membranous lobes with minute spicules apically; there is, on the sides of the segment, a pair of long, slender, pale appendages.

Remarks: The species belongs to the *angustella*-group, like *O. moselyi* which is perhaps its sister species.

Distribution and Ecological Remarks: Described from Lesbos, it was later found on Rhodes and in the Strandzha Mountains of Bulgaria. From the Levant, only one pair is known to the author, caught, with artificial light, on 14 August 1980 on the Litani River (Lebanon) in its lower course at Sir el Gharbii (some 15 km from the river's mouth, ca. 40 m a.s.l.). But it should be added that one female caught in April 1943 at "Hula" (without more information about locality), and kept in the Zoological Museum, Amsterdam, is possibly also *O. melitta*; if the species was once present in Israel, it probably no longer belongs to its fauna.

It would be an error to infer from its presence in the Litani River that the species is a potamobiont: in the Strandzha Mountains it was caught, with artificial light, in very large numbers on a cool water course (Kumanski, 1985).

Orthotrichia costalis (Curtis, 1834)

Figs. 211–218

Hydroptila costalis Curtis, 1834, *Phil. Mag.*, 4:218.
Orthotrichia tetensii Kolbe, 1887, *Ent. Nachr.*, 13:356–359; auct., e.g. Botosaneanu, 1963, *Polskie Pismo ent.*, 33(2):95–96.
Orthotrichia costalis —. Neboiss, 1963, *Beitr. Ent.*, 13(5/6):594–595, etc.

Wing expanse of the only known specimen from Israel: ca. 6 mm. Antennae of this specimen broken; in several publications, the number of antennal segments is given as 35 in the ♂, and 27 in the ♀. Wings described as almost entirely dark brown or blackish. No androconial black scales along the subcosta of the ♂ forewing. Abdominal venter V was described as devoid of the pair of minute warts corresponding to internal glands. In the ♂ a shorter unpaired process posteriorly on venter VI, a longer one covered with very coarse setae on venter VII; such a process only on venter VI in the ♂.

105

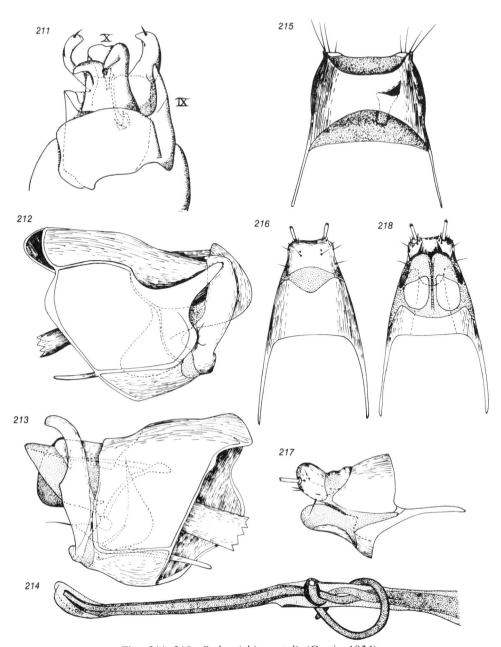

Figs. 211–218: *Orthotrichia costalis* (Curtis, 1834),
specimens from England or from Denmark

211–214. male genitalia (211. dorsal view; 212. lateral view, left side of segments IX & X;
213. lateral view, right side of segments IX & X; 214. distal half of phallic apparatus);
215–218. female genitalia (215. segment VIII, ventral view, with duct and opening of gland;
216. segments IX & X, dorsal view; 217. the same, lateral view; 218. the same, ventral view).
(211: from J.E. Marshall, Trichoptera Hydroptilidae, *Handbooks for the identification of
British insects I*, part 14 (a), 1978; 212–214: from Nielsen, 1957;
215–218: from Nielsen, 1980; some of Nielsen's figures slightly simplified)

Male genitalia: Segment VIII with tergum more protruding than sternum. Segment IX lacking important proximal excisions. Dorsal plate (segment X) moderately long, very asymmetrical, without clear demarcation from segment IX, partly sclerotized and partly membranous; its most distinctive feature is the presence of two strong, sclerotized "hooks" in its distal part: left hook bent upwards, right hook more slender, bent laterally (to the right). In its ventro-distal part, segment IX is devoid of a complex set of appendages; laterally, it has on the left side a very conspicuous horn-like sclerotized appendage, whereas its counterpart on the right side is an inconspicuous bipartite process. Very distinctive is the presence in a lateral position of quite well developed, single-jointed, probably movable inferior appendages; they are asymmetrical: the right one being more slender, especially in its distal part, the left one distinctly broader, distally split into two branches (not seen on the figures!); both gonopods curved, but differently, tips directed mediad. "Bilobed process" only apically split, each short branch with one seta. Phallic apparatus very long and slender; its titillator, strongly coiled around the phallus, is comparatively very short.

Female genitalia: Abdominal segment VII is the longest. The only setae on segment VIII (not numerous, not very long, not curled) are inserted on a pair of narrow sclerotized strips located posteriorly in the middle of each lateral side; the venter is asymmetrical owing to the fact that the terminal part of the duct of the glandular organ, with its opening, is turned to the right. Segment IX with main sclerites whose shape is shown in Figs. 216–218; its distinctive character is the presence on its ventral side of a pair of more or less triangular (or elliptical) sclerites; posterior to these, a pair of large membranous lobes with minute spinules.

Remarks: This species belongs to a small species-group distinct from that which includes the two other Levantine species.

Distribution and Ecological Notes: *O. costalis* is known from many parts of Europe; it has also been mentioned from Tunisia, Egypt, Sudan, Asia Minor, and Iran; its presence is often rather sporadic. Only one specimen from the Levantine Province is known to the author: a male caught on 2 April 1942 on the former marshes of Hula (presently in the Zoological Museum, Amsterdam); I strongly suspect that the species disappeared from Israel together with this habitat; perhaps it will be found somewhere in (northern) Lebanon.

Genus ITHYTRICHIA Eaton, 1873

Trans. ent. Soc. London, 1873: 130 (etc.)

Type Species: *Ithytrichia lamellaris* Eaton, 1873, by original designation and monotypy.

Diagnosis: Small to very small hydroptilids; with ocelli; antennae short in both sexes, with 25 segments at most; postoccipital warts on head large, slightly detached from their substrate but not forming scent-organs; spur formula 0, 3, 4. Wings rather narrow, only moderately pointed, surface very setose and also with erect setae, fringes only moderately long; venation moderately reduced: in forewing SR curiously 4-branched (but only apical fork 2 present!), and M 3-branched (apical fork 3); in hindwing SR 3-branched and M 2-branched (only f2 present). Abdominal sternite V proximally with short setose processes in connection with glands; in both sexes, only abdominal sternite VI with medio-posterior appendage, generally much longer and pointed in the male. Male genitalia with distinctive basic structure, not easy to understand; genital capsule massive, dorsally convex, ventrally flattened; segment IX dorsally entirely "open" (membranous), distally continued by an entirely membranous segment X ("plate") placed above the phallic complex; segment IX also ventrally membranous, but laterally well sclerotized ("side pieces", with anteapical-apical parts diversely formed); always a "subgenital plate", apically with setae; inferior appendages horizontal, elongate, rooted almost at antero-ventral margin of segment IX and extending as much as the most strongly protruding other parts of the genitalia; phallic complex not very long, with constriction slightly distad from its middle, where the rather short titillator, wound once around the phallus, arises. Female abdomen made up of 10, maybe 11 segments, last segments with long apodemes and forming an ovipositor; segment VII the longest, not modified; segment VIII much shorter and narrower, with exceedingly long apodemes, its distal part membranous, ventrally with a row of 6 long setae inserted on small tubercles, venter with species-specific sclerotized formations connected with the opening of the duct of an internal gland; trident-like internal structure in connection with the vagina, well developed.

General Remarks: A very small genus, even if the European "*clavata*" will prove to be a small species complex; there are presently 3–4 western Palaearctic, and 3 Nearctic species (*I. clavata* Morton being reputedly Holarctic; but see remarks on *I. dovporiana*). The genus *Saranganotrichia* Ulmer, with one species in Java, is considered by Marshall (1979:216) as possibly synonymous with *Ithytrichia*. According to Scott (1986:233), the genus is represented in Africa south of the Sahara, too, only young instars being presently known from there. The distribution of most species is (or seems to be) quite restricted, but *I. lamellaris* is widely distributed, as is possibly (?) also *I. clavata*. It is possible that there are two species-groups, one with broader, apically truncate gonopods, the other with slender gonopods tapering apically. The species are generally known as distinctly rheophilous, inhabitants of flowing water — in the Rhithral and the Potamal, exceptionally also in the Crenal — with abundant thickets of higher plants or of mosses; nevertheless, the exact requirements of the different species are unknown. The larvae build extremely characteristic delightful little cases: from secretion only, transparent, very flat, resembling pumpkin seeds, narrow anteriorly with a very small, oval, ridged opening, and a slit at the posterior end (the sealed pupal case being attached to the substrate anteriorly and posteriorly).

Key to the Species of *Ithytrichia* in the Levant

♂♂

1. On dorsal side of genital capsule a pair of parallel, darker "rods"; in lateral view, dorso-proximal part of genital capsule strongly protruding anteriad compared with ventral part; inferior appendages broad, with truncate apices.   **I. lamellaris** Eaton

\- No dark "rods" on dorsal side of genital capsule; in lateral view, ventro-proximal part of genital capsule strongly protruding anteriad compared with dorsal part; inferior appendages slender, tapering towards apices.   **I. dovporiana** Botosaneanu

♀♀

1. Sclerotized plate on venter VIII apically slightly trilobed, sides sinuous, basally broadened.   **I. lamellaris** Eaton

\- On venter VIII a complex of 3 sclerotized formations: median plate followed distally by a conical projection plus one pair of lateral, twisted rods.

     I. dovporiana Botosaneanu

Ithytrichia lamellaris Eaton, 1873

Figs. 219–223

Ithytrichia lamellaris Eaton, 1873, *Trans. ent. Soc. London*, 1873:130, 140–141, 150, Pls. 2 & 3.

Wing expanse in two ♂♂ from Lebanon: 4.5 and 5.1 mm; in one ♀ from Lebanon: 4.9 mm (figures in the literature: 6.3–7.5 mm). One Lebanese ♂ has 23 antennal segments (22–25 segments for the ♂, and 19–22 for the ♀ in published information). In both sexes a rather long, pointed, median appendage at the posterior limit of venter VI.

Male genitalia: Genital capsule in lateral view with narrow (= low) dorsal part strongly protruding anteriad compared to the proximally rounded ventral part from which it is separated by a moderately deep but low triangular excision; in dorsal view, central parts of segment IX membranous, but with a pair of very characteristic, parallel, longitudinal, dark stripes (or "rods"; more correctly: low keels); the sclerotized "side pieces" of segment IX are, in dorsal view, rather broad basally, their dark, slightly capitate apices turned laterad and downwards. Segment X reduced to a membranous formation above the phallic complex (Fig. 220). Just below the phallus, a distinctly bilobed "subgenital plate", each lobe with a bristle (Figs. 219, 221). At a still lower level, a membranous "bilobed process" (Fig. 220). Inferior appendages strong, horizontal, more protruding than the other parts of the genitalia; seen dorso-ventrally, broad, almost parallel, apices obliquely truncate; in lateral view broad basally, gradually narrowing to a blunt apex, before which is a characteristic, laterally placed small "comb" made up of several black tubercles (seen, at slight magnifications, as black spot). Phallic complex comparatively short and stout, apical part distinctly separated from the rest, apex broadened; darker apical part of ductus ejaculatorius broadened; titillator comparatively short, wound once around the phallus.

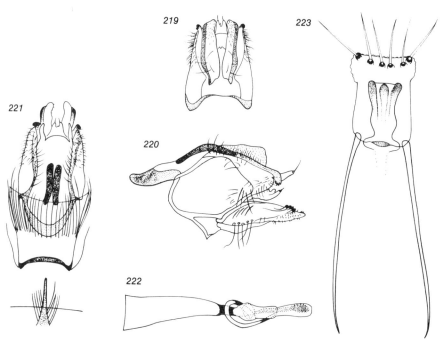

Figs. 219–223: *Ithytrichia lamellaris* Eaton, 1873
219–222. male genitalia, specimens from Romania or from Bulgaria
(219. dorsal view; 220. lateral view; 221. ventral view;
222. phallic apparatus, lateral view);
223. segment VIII of female, ventral view.
(219–221: from L. Botosaneanu & E.A. Schneider,
Studii Comun. Muz. Brukenthal, St. Nat., 22, 1978;
222: from Kumanski, 1985)

Female genitalia forming an ovipositor. Segment VII very long; segment VIII much shorter and narrower, (partly) invaginated in segment VII, with extremely long apodemes; posterior part membranous, with ventral row of 6 strong setae arising from small rounded projections; in the middle of the venter a sclerotized plate (corresponding to the terminal part of the duct of an internal gland), of characteristic shape: distally trilobed (lobes similar), sides sinuous, much broadened basally.

Distribution and Ecological Remarks: *I. lamellaris* was found in most parts of Europe and also in Asia Minor. In Lebanon it is presently known, mostly from young instars, from some coastal basins (main course of Nahr ed Damour; several tributaries of Nahr el Aouali, from 850 to ca. 40 m a.s.l.), and also from the Orontes at Hermel (650 m a.s.l.); it probably inhabits sluggish reaches of these water courses, with abundant vegetal growth. In Israel, the only known locality is the complex of springs and streams at Tel Dan (Tel el Kadi) in Upper Galilee (1), which is almost certainly the southernmost point in the species' distribution. The few adults were caught, in the Levant, either in May, or in September–October.

Ithytrichia dovporiana Botosaneanu, 1980
Figs. 224–234

Ithytrichia dovporiana Botosaneanu, 1980, *Bull. zool. Mus. Amsterdam,* 7(8):74–75, Figs. 1
E–H. Type Locality: Naḥal 'Arugot, in the Dead Sea Depression, Israel. Type deposited in
the Zoological Museum, University of Amsterdam.

Wing expanse: ♂ 4.1–4.6 mm (one specimen measured 3.65 mm), ♀ 3.8–4.5 mm.
Antennae extremely short in both sexes, with 23 segments in most of examined ♂
specimens (24 in one of them; 21 mentioned for the holotype, a "metamorphotype");
in the ♀ with 20–21 segments. In both sexes large postoccipital warts on head, slightly
detached from substrate but apparently not forming, or covering, scent-organs. Both
wings covered with dense reddish- brown setae, basally in the costal-subcostal zone
of forewing many black, erect setae, fringes not very long in forewing, and not
exceedingly long in hindwing.

Male genitalia: Genital capsule in lateral view proximally with dorsal part low and
very short compared to the massive and proximally somewhat irregular ventral part
from which it is separated by a moderately deep excision forming an acute angle (the
sclerotized ridge proximally bordering the dorsal part obliquely descending
posteriad). In dorsal view, central parts of segment IX membranous, without dark
longitudinal "rods", distally directly continued by membrane probably belonging to
the segment X, the central part of which is a large, rounded plate (in lateral view
resembling a strong pointed projection). The sclerotized "side pieces" of segment IX
("a" in the figures) are, in dorsal view, rather narrow overall, with parallel sides; there
is a distinct zone covered with minute spinules on each side; the distal parts of the
"side pieces" are very characteristically shaped — see Figs. 224 and 228: their
preapical, more or less obtuse part is suddenly followed by a quite distinct bend
separating it from a moderately long digitiform projection ("b") with obliquely
truncate or somewhat rounded apex, more dorsally and more medially placed than
the preapical part of the "side piece", and distinctly protruding beyond it (the two
digitiform projections are divergent); on ventral and median side of each "side piece",
just basally from the mentioned bend, a hump with one spine ("c"). Just below the
phallus, the subgenital plate ("e") with broad proximal part and narrow, conical,
distinctly elongated, not bilobed distal part terminating in two setae. Inferior
appendages ("d") horizontal, slender, slightly divergent, tapering distad but without
pointed apices; these apices blackened, with several obtuse tubercles, distinctly bent
dorsad and mediad; anteapically with an obtuse median projection with two spines.
Phallic complex with apical part not separated from the rest; around the preapical part
of ductus ejaculatorius an elongated, darker "muff"; titillator comparatively short,
wound once around the phallus.

Female genitalia: Shape and proportions of last segments as in *I. lamellaris.* In the
middle of the venter of segment VIII, a characteristic complex of sclerotized
formations certainly connected with an internal gland; just in the middle there is a

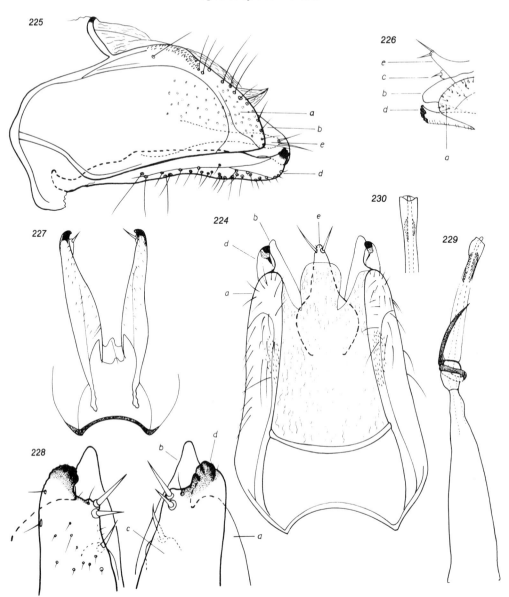

Figs. 224–230: *Ithytrichia dovporiana* Botosaneanu, 1980, male genitalia
224. dorsal view; 225. lateral view;
226. apical parts of genitalia, lateral view, seen on opposite side;
227. inferior appendages, ventral view; 228. apex of the two inferior appendages,
with side-pieces of segment IX, ventral view, strongly magnified;
229. phallic apparatus, lateral view; 230. apex of phallic apparatus, dorsal view
In various figures: a – "side pieces" of segment IX; b – apical projection of these
"side pieces'; c – hump on ventral and median side of these "side pieces";
d – inferior appendage; e – "subgenital plate". (226, 227, 230: from Botosaneanu, 1980)

112

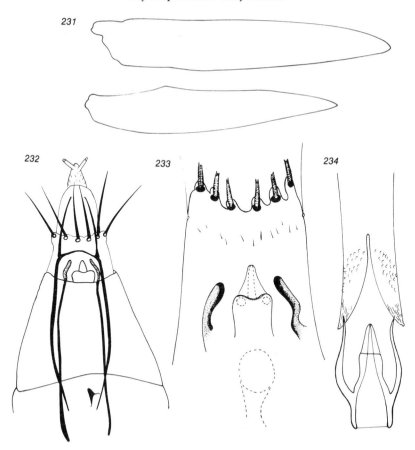

Figs. 231–234: *Ithytrichia dovporiana* Botosaneanu, 1980
231. wings, venation not represented;
232–234. female genitalia
(232. ventral view; 233. segment VIII, ventral view, strongly magnified;
234. trident-like formation and vagina)

short plate with rounded posterior angles and emarginated distal margin where a rather long, conical formation is inserted; proximal parts (if any?) of this plate quite indistinct; on each side of the plate, and at some distance from it, a twisted, dark "rod", the two rods (representing the distal ends of the apodemes VIII) with apices converging. The trident-like internal structure with very slender lateral arms which are only very slightly longer than the median part of the trident; on the microscopical preparations, the vagina (with which this trident-like structure is connected) is always perfectly distinct (Fig. 234), proximal part of its walls finely but distinctly sculptured.
Remarks: *I. dovporiana* was initially described from a male "metamorphotype", this description being incomplete and with some interpretation errors. The species seems to be closely related to *I. clavata* Morton, 1905; this latter species was described from

113

New York State, subsequently reported from several other states of the U.S.A. and from Canada, and — since 1930 — also from a few western Palaearctic countries: Sweden, Finland, the British Isles, France (Pyrenees), Morocco (Middle Atlas), and Tunisia. Some of the western Palaearctic records are accompanied by illustrations, others not, and there is also one illustration lacking a mention of a locality.

Thorough comparison of *I. dovporiana* with North American specimens of *I. clavata* (from Canada: Quebec, Ottawa River, Norway Bay, 4 Aug. 1973, leg. O.S. Flint, Jr — specimens of both sexes kindly sent to the author by Dr O.S. Flint, Jr) reveals beyond any doubt that the two species are perfectly distinct. The most important genital characters of *I. clavata* distinguishing it from *I. dovporiana* are the following. In the male: "subgenital plate" much shorter (stouter); "side pieces" of segment IX without any preapical bend, regularly tapering to a very slightly enlarged apex with minute "beak" directed laterad; zones of dorsum IX covered with minute spines, larger; medio-ventral setose humps of "side pieces" much more protruding; gonopods tapering to blunt apices which are less blackened (only a distinct black "point" in the outer- and dorsal-apical corner), not bent mediad and dorsad, devoid of anteapical setiferous hump. In the female: sclerotized plate on venter VIII longer, its proximal (internal) parts complex and well sclerotized, apical excision deeper, with a shorter, rhombic formation inserted in it . Most of these characters of *I. clavata* were well described and illustrated by Ross (1944: 124, figs. 457 and 459), who is probably the only author to have expressed doubts about the correctness of the attribution of European specimens to *I. clavata*. For obvious reasons, the problem of the correct attribution of these specimens (and of those from North Africa) cannot be tackled here. I hope to deal with it in another publication; a small complex of species might be involved here, *I. clavata* being possibly one of them.

Distribution and Ecological Remarks: *I. dovporiana* is presently known only from Naḥal ʿArugot, one of the two major water courses at ʿEn Gedi, on the western shores of the Dead Sea, Israel, at 350 m b.s.l.(13), being quite possibly an endemic element of the Levantine province, restricted in its distribution to the desert zones. It is interesting to note that the species was never caught (neither as adults nor as young instars) outside Naḥal ʿArugot, although it was searched for with equal diligence in other potentially suitable localities in the Dead Sea Depression and in the Negev, including the neighbouring, and apparently very similar, Naḥal Dawid. The type was caught on 4 May 1970; on 26 April 1984 I caught a good series by artificial light (11 males, 5 females: topotypes). The population is a large one, the characteristic larvae and pupae always being observed in large numbers on various substrates in this desert stream fed by springs.

Family PHILOPOTAMIDAE Stephens, 1829

Cat. Brit. Ins., 1:316

Type Genus: *Philopotamus* Stephens, 1829.

Small or medium sized insects. Ocelli present. Antennae stout, with short segments, scapus only slightly stronger. Maxillary palpi in both sexes with 5 segments, 1st short, 2nd with brush of stiff setae at medio-distal angle, last segment long, annulate, flexible, as long as 3rd and 4th together. Spurs: mostly 1, 4, 4 or 2, 4, 4. Forewings oval, apex rounded, with f 1–5 or 1, 2, 3, 5, discoidal and median cells closed. Hindwings shorter, as broad or slightly broader, f 1, 2, 3, 5, discoidal cell usually closed and median cell open. Tibiae and tarsi of intermediate legs of female sometimes slightly dilated. Male genitalia simple, with long and strong gonopods, frequently bisegmented. Female sometimes with, and sometimes without, short ovipositor.

Key to the Genera of Philopotamidae in the Levant

1. Spurs 1, 4, 4. In maxillary palpi, 1st segment short, 2nd long. Intermediate legs in female slightly dilated. Inferior appendages of male unisegmented. **Chimarra** Stephens
– Spurs 2, 4, 4. In maxillary palpi, 2nd segment short. Intermediate legs in female not dilated. Inferior appendages of male with coxopodite and harpago.

Wormaldia McLachlan

Genus CHIMARRA Stephens, 1829

Cat. Br. Ins., 1:318

Type Species: *Phryganea marginata* L. (by monotypy).

Diagnosis: Small to medium-sized species (Fig. 235), variously and sometimes conspicuously coloured, wings only with slight pubescence, leaving venation distinct. Antennae shorter than forewings. Maxillary palpus long, thick, 2nd segment long and bearing a tuft of bristle-like setae at apex. Wings moderately elongated and slender, hindwings slightly broader than forewings; several features of the venation are more or less typical; for the forewings: forks 1, 2, 3, 5; discoidal, medial, thyridial cells present; radius thickened, sinuous in its middle part; SR sinuous before the discoidal cell, thickened at the base of this cell; between the bases of the discoidal and medial cells there is a glabrous, shining space (the "nude cell"); in the hindwings: forks 1, 2, 3, 5 (1 very narrow); a very characteristic feature are two anal veins (A2 and A3, or A1 and A2 — depending on the interpretation) which together form a kind of closed cell, or loop, near the wing's base. Spur formula: 1, 4, 4 (the anterior spur can be very small); in the female, tibiae and tarsi of intermediate legs slightly dilated. Male genitalia only slightly protruding, abdominal sternites VIII and IX (sometimes also VII) with median, elongated processes; claspers unisegmented. Last abdominal segments in female forming a short, stout complex; segment VIII fused in a cylinder whose apical margin bears setose tubercles.

General Remarks: *Chimarra* comprises a very large number of species, almost all occurring in tropical-subtropical (and also warm temperate) zones, the Oriental, Ethiopian, and Neotropical Regions being the most productive. In contrast, there are only 17 Nearctic species, whereas Europe and Palaearctic Africa have only one. It is therefore not surprising that the Levant also has a single species. Unfortunately, the affinities within the genus are very poorly known, and there is no monographic revision. It is, nevertheless, recognized that there is extreme diversity, especially with regard to the male genitalia. Generally speaking, the species are inhabitants of warmer running (not fast running) water. The larvae are typical retreat-makers.

Chimarra lejea Mosely, 1948

Figs. 235–241

Chimarra lejea Mosely, 1948, *British Museum Expedition to South-West Arabia 1937–8*, 1:75–76, Figs. 17–22.

Wing expanse, in the Levantine specimens: ♂ 9.3–10.5 mm, ♀ 10.6–11.6 mm (considerably larger in specimens from Yemen: length of anterior wing in both sexes 6 mm, giving an expanse of ca. 18 mm). Habitus: Fig. 235. Head yellow, clothed with golden-yellow setae; antennae dark ochreous, with no very marked annulations; maxillary palpi dark ochreous; wings vitreous, with a very light, fuscous vestiture; venation typical for *Chimarra*: Fig. 236; legs and abdomen ochreous.

Male genitalia: From the centre of the margin of venter VIII arises a single process, much shorter than that of venter IX, with the shape shown in Figs. 238 and 239 (these median processes are symmetrical, not as represented in Fig. 239). Central portion of dorsum X (described as dorsum IX by Mosely) membranous; on each side of this membranous area, a slender sclerotized process with apex clubbed and laterally turned in dorsal view, obliquely truncate when viewed from the side. Latero-apical angles of dorsum IX produced, setose. Superior appendages (cerci) small, rounded, lying near bases of the lateral processes of dorsum X. A very long ventral process, extending to the apex of the inferior appendages, arises from the centre of the distal margin of sternite IX; this process regularly broadens to an emarginated apex. Inferior appendages one-segmented, in the shape of broad, oblong plates arising from narrower bases, apices obliquely truncate, dorso-distal angle slightly inturned. Phallic apparatus revolver-shaped, with an endothecal armature represented by two very long, fine, black, asymmetrical spines. Beneath it is a "lower penis cover" appearing in lateral view as down-turned hook.

Female genitalia forming a stout, short ovipositor. Medio-ventrally on segment VII a rather well developed keel. Segment VIII much smaller, with distal margin as in Fig. 241 (in this figure, neither the boundary between segments IX and X nor the cerci of this last segment are shown).

Remarks: It is believed that this species is related to *C. clara* Mosely, from Ruwenzori, Uganda.

116

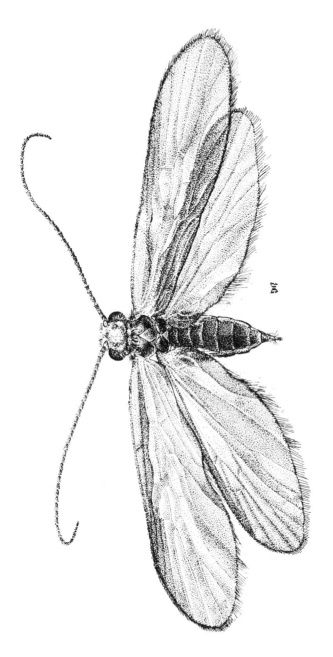

Fig. 235: *Chimarra lejea* Mosely, 1948, female

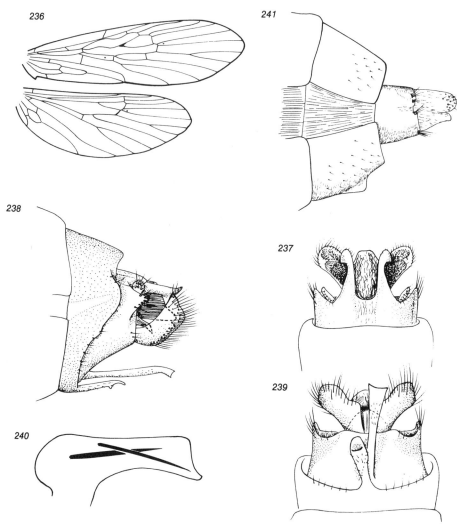

Figs. 236–241: *Chimarra lejea* Mosely, 1948
236. wings;
237–240. male genitalia
(237. dorsal view; 238. lateral view; 239. ventral view;
240. phallic apparatus, lateral view, diagrammatic);
241. female genitalia, lateral view.
(From Mosely, 1948)

Distribution and Ecological Notes: *C. lejea* was first described from the south-western corner of the Arabian Peninsula: two localities — waterfalls — in the former Western Aden Protectorate, and one wadi in Yemen, near Ta'izz (all at 1,900–2,100 m a.s.l.). It was rediscovered in a warm stream at Wondo Abella, Ethiopia. No wonder that the species is restricted, in the Levant, to the typical water courses of the desert zones. Three localities are presently known from the western shores of the Dead Sea (13), at some 350 m below sea level: the complex of springs and streamlets at 'Ein Fashkha, possibly the northernmost point in the species' distribution, where an important population is present; the springs and streamlets at 'Ein Turabe, with an apparently important population; and Naḥal Dawid, near 'En Gedi, where several females were caught. Moreover, there is one known locality in the 'Arava Valley (14): Ne'ot haKikar, at 350 m below sea level (1 male known). It is somewhat surprising that it has not yet been caught in other, apparently suitable localities in the Negev and in the Sinai; it is certainly present in Jordan. The Levantine specimens were caught mostly from April to June (1 female caught in October).

Genus WORMALDIA McLachlan, 1865
Trans. ent. Soc. London, (3) 5:140

Type Species: *Hydropsyche occipitalis* Pictet, 1834 (subsequent selection by Ross, 1949).

Diagnosis: This is an extremely polymorphic genus (see especially: Ross, 1956:38–46, 61–76; more species were subsequently discovered), and it would be not only difficult, but also pointless for the purposes of this Fauna volume, to give a general diagnosis. The following will be merely a diagnosis of the western Palaearctic species-group *occipitalis*, to which the Levantine species belongs. Medium-sized insects, with monocolorous (mostly brownish or greyish) wings, forewings with distinct pale zones at the anastomosis, at the base of median cell, and at arculus. Second segment of maxillary palpus short. Spurs 2, 4, 4; intermediate tibia and tarsi in the female not dilated. In both wings, forks 1, 2, 3, 5. No mesal processes on abdominal sternites. Male genitalia simple (excepting the endothecal armature of the phallus); tergite VIII not produced over base of genitalia, generally with medio-distal emargination; cerci digitiform, attached at base only; segment X triangular (ogival), without dorsal, apical, or lateral "flaps" ("flanges"), but generally with a simple, preapical dorsal point (the apex not being entirely recurved to form a hook directed backwards); inferior appendages simple, bisegmented, 2nd segment more or less elongated; the membranous endotheca inside the phallic apparatus with complex set of sclerotized spines, etc., excellent for enabling distinction of species or subspecies. Last segments of female abdomen forming a moderately long ovipositor, segments gradually becoming smaller, telescoped, segment VIII small, IX + X even smaller, VIII and IX with very conspicuous apodemes.

General Remarks: *Wormaldia* is a genus of medium scope; some 40 species were known to Ross (1956), with some more known at present; they are Palaearctic, Nearctic, Oriental, Ethiopian, and Neotropical. The European fauna (*s. lat.!*) presently comprises about 20 species, some with geographical races, most of them distributed in the perimediterranean zones, many of them belonging to the *occipitalis*-group.

Wormaldia subnigra McLachlan, 1865

Figs. 242–246

Wormaldia subnigra McLachlan, 1865, *Trans. ent. Soc. London*, (3) 5:142, Pl. 13, figs. 24–25.
Botosaneanu, 1982, *Bull. zool. Mus. Amsterdam*, 8(22):178, Fig. 40.

Wing expanse in Levantine ♂ specimens: ca. 13.5–14.6 mm (in ♀ sometimes slightly larger). Body blackish, but pronotum and warts on head yellow-brown; antennae dark brown with narrow yellow annulations; forewing grey with short, uniformly blackish-brown vestiture, some distinct pale spots, and dark venation (Fig. 242). Second segment of maxillary palpus only as long as the first. In both wings, forks 1, 2, 3, 5; no "nude cell" in the forewing.

Male genitalia: Tergite VIII with deep, rounded, medio-distal excision; sternite VIII with less deep, rounded, medio-proximal emargination; IXth sternite with very deep medio-proximal triangular excision; in lateral view, segment IX stout, much longer ventrally than dorsally, with oblique posterior margin. Cerci (superior appendages) long, slightly sinuous in lateral view, with obtuse apex slightly downturned; in dorsal view they are characteristically shaped: slender, apices slightly capitate with distinct point bent mediad. Segment X slightly longer than cerci, more or less ogival but with spatulate apical part having a strong preapical point turned upwards and forwards (very conspicuous in lateral view). Inferior appendages very long compared with the rest of the genitalia, with common root, bisegmented, proximal segment much stouter and slighty longer than distal segment which is slightly curved, tapering gradually to an obtuse apex having on its median side a distinct black "button". The very well-developed endotheca (internal membranous sack) of the phallic apparatus contains a typical set of sclerotized formations which are (from distal to proximal): a small cone consisting of numerous fine, agglutinated spinules; a pair of large spurs; a pair of rather large, distinctly pectinate formations; one or two small bunches of fine spines.

Female genitalia: Each of the last abdominal segments very distinctly smaller than the preceding one; they form a relatively long ovipositor. Segment VIII posteriorly truncate, retracted into segment VII. Segments VIII and IX each with a pair of conspicuous apodemes. Segments IX + X firmly united to form a small body retracted into segment VIII.

Remarks: *W. subnigra* belongs to the entirely European species-group of *occipitalis*. Despite its large distribution, the species has no interesting geographic — or other — variability, even in the set of endothecal formations, quite unlike other species of the same group: there is no sensible difference between Levantine and other specimens.

Distribution and Ecological Notes: This species is largely distributed in Europe, including some Mediterranean islands, and is also known from Asia Minor. It is on the list of endangered species in the countries of the European Community. In the Levant, it was found only in Lebanon, where the southernmost limit of its distribution is reached. Most surprisingly, it inhabits only the small coastal drainage basins: no

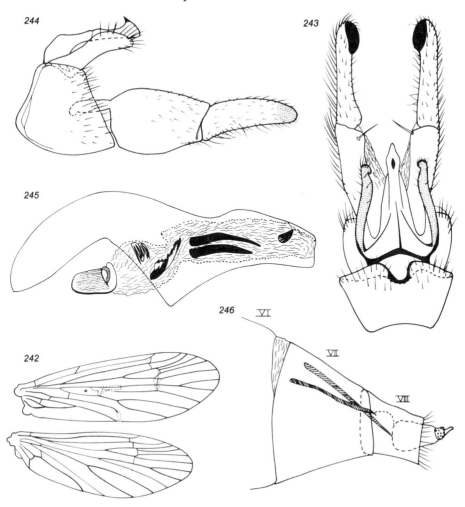

Figs. 242–246: *Wormaldia subnigra* McLachlan, 1865
242. wings;
243–245. male genitalia
(243. dorsal view; 244. lateral view; 245. phallic apparatus);
246. female genitalia

specimen was taken in the — presently well explored — water courses of the Orontes or Litani catchments. The known localities are rather numerous in the basins of the Nahr ed Damour and Nahr el Aouali, these being streamlets and larger streams (perhaps also some springs), mostly at altitudes of 700–1,000 m a.s.l. (Aouali), but also at much lower altitudes in the basin of the Damour (40–260 m). In Lebanon it is a typical rhithrobiont, often forming considerable populations. The species was found on the wing from April to December, with apparently a peak of imaginal emergence in September–November.

Family HYDROPSYCHIDAE Curtis, 1835
Br. Ent., Pl. 544 (text)

Type Genus: *Hydropsyche* Pictet, 1834.

This diagnosis is valid for the subfamily represented in the Levant: Hydropsychinae.

Medium sized to rather large insects, sometimes smaller. No ocelli. Maxillary palpi with 5 segments in both sexes, 1st very short, 5th very long, annulate, and flexible. Antennae very slender, somewhat longer than forewing (especially in male), scapus globose, following segments elongate. Spurs mainly 2, 4, 4. External claw of anterior legs in male hidden in a conspicuous tuft of black setae. In female tibiae and tarsi of intermediate legs dilated. Forewings narrow, distally rather angular (truncate), covered by dense but short setae sometimes forming distinct patterns; venation very complete, discoidal and median cells closed, f 1–5 present. The hindwings sometimes moderately ample, not pointed, their discoidal cell closed, f 1, 2, 3, 5 (or 2, 3, 5) present. Abdominal sternite V without long filiform appendages. Male genitalia very simple (excepting sometimes the phallic apparatus), gonopods slender, bisegmented. Female without ovipositor.

Key to the Genera of Hydropsychidae in the Levant

1. More robust insects. Segments in basal part of antennae with dark, obliquely transverse lines (less evident in freshly emerged insects!). In hindwing median cell closed.
    Hydropsyche Pictet
–   More slender insects. Antennal segments without dark lines. In hindwing median cell open. **Cheumatopsyche** Wallengren

Genus CHEUMATOPSYCHE Wallengren, 1891
K. svenska VetenskAkad. Handl., 24(10):138, 142

Type Species: *Hydropsyche lepida* Pictet, 1834 (by monotypy).

Diagnosis: Hydropsychids generally smaller than *Hydropsyche* and of more slender appearance; forewings generally without conspicuous patterns, covered by short, fine, greyish or brownish setae. Segments of the basal part of the antennae without oblique black lines. In the maxillary palpi, 3rd segment as long as, or longer than 4th. Median cell in hindwings always open. Spur formula: 2, 4, 4. Very simple male genitalia, tergite IX posteriorly with two lateral projections with very long setae, laterally to tergite X a pair of elongated appendages, phallic apparatus distinctly inflated (bulbous) basally, apically with 2 large lateral "scales". As in *Hydropsyche*, tibiae and tarsi of median legs of female dilated. Female genitalia as in *Hydropsyche*.

122

General Remarks: *Cheumatopsyche* is a medium-sized genus, well represented in Asia (some 30 known species), North America (some 40), and especially Africa (some 50), with only a very small number of species (maybe 3) in the western Palaearctic. Chaos still reigns in the systematic knowledge of the genus in some parts of the world, and distinguishing related species from one another is often an arduous task, particularly in view of the very simple genitalia. There is apparently only one species in the Levantine Province, but its true nature is difficult to define (see below).

Cheumatopsyche capitella (Martynov, 1927)
[*C. capitella* (Martynov, 1927) x *C. lepida* (Pictet, 1834)]?
Figs. 247–269

Bibliographic information: see under Remarks.

Wing expanse (Levantine specimens): ♂ 10.1–10.8 mm, ♀ 11.8–12.3 mm. Head and body brown, legs paler, antennae paler, too, with brown annulations, wing membrane greyish-iridescent, vesture dense but very short and fine, golden and brownish-reddish patches alternating on forewing to give a faint, regular pattern (but the forewings may also be dark grey). Wing venation: Fig. 266.

Male genitalia: Laterally on posterior margin of tergite IX are located the two customary rounded projections with long, strong setae; between them, the margin has a clearly marked, short median projection (Figs. 248, 258), sometimes lacking (or, more correctly: disappearing if the genitalia are viewed at a slightly different angle, the margin between the lateral projections then appearing almost straight, with a minute central indentation: Figs. 247, 257). Tergite X in dorsal view triangular (Figs. 247, 248, 257), but becoming more rounded if viewed in a slightly different position (Fig. 258); in lateral view (Figs. 249, 250, 259) almost always with distal border sinuous, preapically more or less depressed. The two projections laterally accompanying tergite X are as long as the tergite in lateral view (Figs. 249, 250, 259, 260), and almost always slightly narrowed before the apex which is generally clearly swollen (capitate); the setose warts near the roots of these projections always strongly developed. Second segment (harpago) of inferior appendages never really claw shaped, but of normal shape although rather slender, with two minute spines at tip. Phallic apparatus: the medio-ventral keel hanging from the preapical part is sometimes well developed (Figs. 256, 264), sometimes much less so (Figs. 255, 265), never extremely developed.

Female genitalia: An interesting feature is the heavily sclerotized "Zangengrube" (= recept. harp.) on segment IX (Tobias, 1972), a formation receiving the harpago during copula (black in Figs. 267–269). This is very distinctly capitate at its dorsal end, and its distal border ("marg. dors. foram.") is not sinuous: this is a clear difference from *C. lepida*, as properly described and illustrated by Tobias (1972:399, fig. 176), but, unfortunately, there is no description of the female genitalia of *C. capitella*.

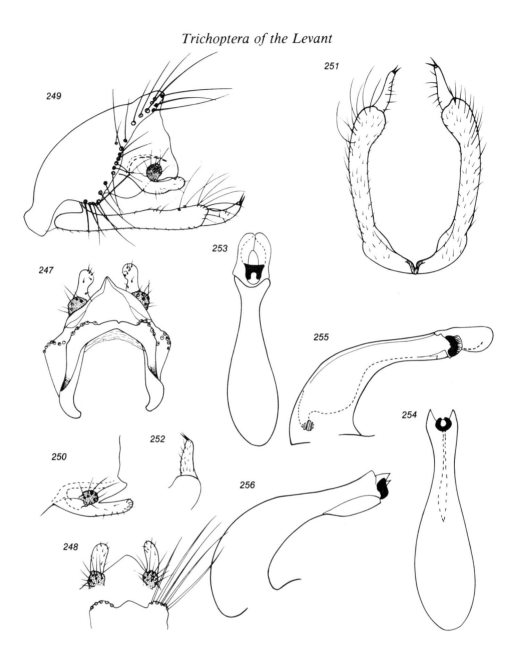

Figs. 247–256: *Cheumatopsyche capitella* (Martynov, 1927)
or *C. capitella* x *C. lepida* (Pictet, 1834), male genitalia
(247. dorsal view; 248. segment X and distal limit of segment IX, dorsal view;
249. lateral view; 250. segment X only, lateral view;
251. gonopods, ventral view; 252. apex of gonopod, median side;
253–254. phallic apparatus, dorsal view, in 254 without its apical lobes;
255–256. phallic apparatus, lateral view, in 256 without its apical lobes).
247, 249, 251, 253, 255: specimen from Nahr ed Damour, Lebanon;
248, 250, 252, 254, 256: specimen from Tanuriia, Golan Heights

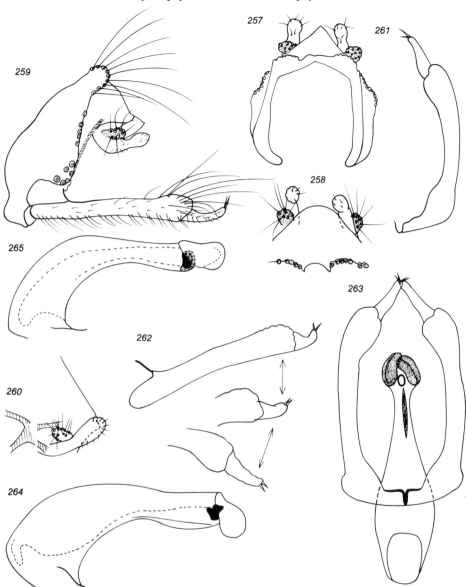

Figs. 257–265: *Cheumatopsyche capitella* (Martynov, 1927)
or *C. capitella* x *C. lepida* (Pictet, 1834), male genitalia
(257. dorsal view; 258. segment X and distal limit of segment IX,
dorsal view, same specimen but different angle; 259. lateral view;
260. segment X only, lateral view; 261. gonopod, dorsal view;
262. gonopod, lateral view, with apical part seen from three different angles;
263. gonopods and phallic apparatus, ventral view;
264–265. phallic apparatus, lateral view).
257, 258, 259, 261, 265: specimen from Litani, Lebanon;
260, 262, 263, 264: specimen from Dan, Israel

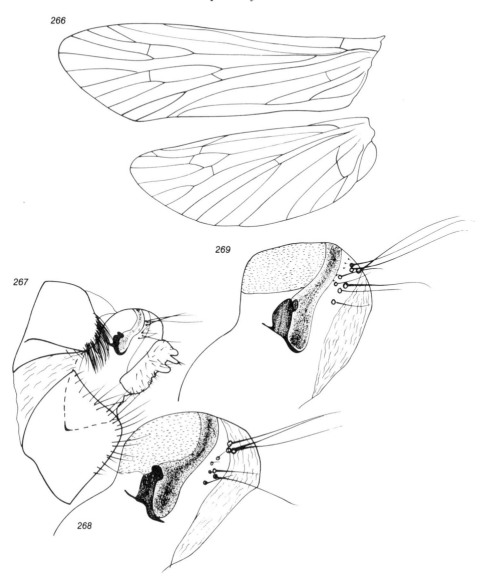

Figs. 266–269: *Cheumatopsyche capitella* (Martynov, 1927)
or *C. capitella* x *C. lepida* (Pictet, 1834), male genitalia
266. wings;
267–268. female genitalia, lateral view,
and segment IX more strongly magnified,
specimen from Litani, Lebanon;
269. segment IX of female, lateral view,
specimen from Nahr ed Damour, Lebanon

Remarks: Two species were mentioned from the Levant in various publications:
1. *C. capitella* (Martynov, 1927) (Martynov, 1927a: 186–188, Pl. X, figs. 45–47); to this day this is the only description of the species, aside from a shorter Russian translation in Martynov (1934: 284–286). *C. capitella* was described from several localities in Soviet Central Asia; it was subsequently mentioned, unfortunately without description or illustration, again from Central Asia, and also from eastern Asia Minor, north and south Iran, Afghanistan, Mongolia, as well as from Greece and from the United Arab Emirates. It was mentioned in several publications from Israel and from Lebanon.
2. *C. lepida* (Pictet, 1834), described as *Hydropsyche lepida* (*Rech. Phryg.*, p. 207, Pl. 18, fig. 1). Subsequently redescribed and illustrated several times (best existing illustration in Tobias, 1972: Figs. 171–176). This species is known from almost everywhere in Europe; it is simply mentioned also from Asia Minor, North Africa (although substituted there by what seems to be a closely related species), northern Iran, and the "Levant" (Malicky & Sipahiler, 1984: 209).

If all available information is considered, the two species should coexist in eastern Asia Minor, northern Iran, and in the Levant. If these are really distinct species, the males differ only in quite minute characters, as already shown by Martynov (1927a, 1934). I have no reason to suspect that more than one species is present in Lebanon and Israel. However, after careful study of the few available specimens, I feel unable to reach a definite decision about their specific attribution. Two male genital characters seem to indicate that *C. capitella* is involved: lateral projections of tergite X clearly capitate and preapically narrowed; and 2nd segment of gonopods never really claw-shaped, although slender. On the other hand, the distinct median projection of the distal margin of tergite IX is considered as characteristic for *C. lepida*. The ventro-distal keel on the phallus is never as strongly developed as in *C. lepida*, but sometimes even more attenuated than described for *C. capitella*. Moreover, the "recept. harp." of the female segment IX is without any doubt different from that of *C. lepida*. Such problems were certainly not unknown to Martynov, and there is one interesting – although never mentioned – sentence in his 1934 book (p. 288) concerning *C. lepida*: "The differences from *Ch. capitella* Mart. become less clear southwards". All this makes me strongly suspect that some introgression is involved in the vast overlapping distribution areas of the two taxa.

Distribution in the Levant and Ecological Remarks: *Cheumatopsyche* was caught in Lebanon in the lower reach of Nahr ed Damour, and in the Litani at Sir el Gharbii (less than 200 m a.s.l.). One unsatisfactorily labelled male specimen comes possibly from Tannuria, in the Golan Heights (18), at 575 m a.s.l.; Dr Heather J. Bromley (Jerusalem) informs me that this locality is otherwise known as the Nahal Gamla springs, draining into the north-east "corner" of Lake Kinneret, and that *Cheumatopsyche* larvae were collected from several localities along this stream. In Israel, the known localities are: Huliot (Sedé Nehemya) at the confluence of the water courses forming the Jordan (1); Dan — one of these water courses (1); an aqueduct with rapidly flowing, turbid and eurythermic water at Bet She'an (7); Buteicha (7);

and the species was also caught in the "Jordan Park" (7); these localities are from ca. 200 m a.s.l. to ca. 200 m below sea level. The number of specimens caught was always surprisingly low; with artificial light, females may be attracted in larger numbers than males. They were found on the wing from April to October. The species probably inhabits only water courses belonging to the Metarhithral, Hyporhithral, and Epipotamal. I suspect that earlier it also belonged to the fauna of Jordan River, and that it is present in the Orontes, although it has not yet been caught from this river.

Genus HYDROPSYCHE Pictet, 1834
Rech. Phryg., p. 199

Type Species: *Hydropsyche cinerea* Pictet, 1834, a synonym of *H. instabilis* (Curtis, 1834); subsequent selection by Ross, 1944.

Diagnosis: Large or medium sized caddisflies (Fig. 270) with a moderately slender appearance even in large specimens, especially when the wings are folded. Colour of wings and body paler or darker in the different species; forewings covered with dense, short setae forming more or less distinct patterns (unfortunately variable and difficult to describe and to use as distinctive characters for species). Antennae very slender, longer than the wings; in their basal part, starting with the 3rd segment, there is always a very distinctive oblique black line on each segment. Maxillary palpi with 3rd and 4th segments triangular, the former slightly shorter than the latter. Spurs: 2, 4, 4. In the male, the external claw on all legs is characteristically hidden by a tuft of black setae; in the female, intermediate tibiae and tarsi dilated. Anterior wings narrow, obliquely truncate at apex; discoidal, median, and thyridial cells all closed, forks 1–5 present. Posterior wings much shorter, broader, folded, obtuse at apex, with forks 1, 2, 3, 5 and with closed median cell. On some abdominal segments, vestigial branchial filaments. Male genitalia (see Fig. 271 for terminology used in the present keys and descriptions): segment IX much higher than segment X with a large, rounded, setose distal projection in its ventral half; boundary between segments IX and X sometimes indistinct, but marked by a row of long setae; a medio-dorsal keel on segment IX; in many species, each of these segments has a dorso-lateral depression; segment X more or less protruding dorso-apically, with triangular, lateral spiny zones (sometimes interpreted as superior appendages), in some species with a pair of apical digitiform appendages; inferior appendages long, slender, bisegmented, harpago always shorter than coxopodite; in the Levantine species the phallic apparatus is generally quite simple, at most with a pair of lateral preapical projections; nevertheless, if carefully observed, its lateral and dorso-ventral outlines offer good distinctive characters for the different species. Abdominal apex in the female very stout; segment VIII with tergite deeply excised medially, and with sternite completely split into two valves; segment IX (see Fig. 272 for terminology used in the present keys and descriptions) offering the most reliable distinctive characters (but possibly not between very closely related species); the conformation of its different parts in each species strictly fits that of the male genitalia and especially of the inferior appendages; segment X small, with 3 dissimilar pairs of cerci.

128

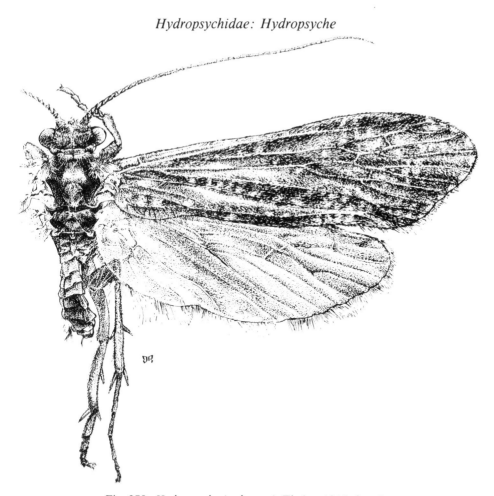

Fig. 270: *Hydropsyche jordanensis* Tjeder, 1946, female

General Remarks: *Hydropsyche* is a large genus with some 200 known species, most of them in the temperate parts of the world (some 90 in the Palaearctic, some 70 in the Nearctic), the Ethiopian and Oriental fauna being distinctly less rich, and no species being known from the Neotropical region. All the species are inhabitants of running water and especially of large, or the largest, rivers; as such, they are endangered in many parts of the world, and several cases of recently extinct species have been reported. The populations are in some cases enormous, and many *Hydropsyche* are excellent indicator elements of water quality or for biological zonation of running water. The larvae build irregular shelters from mineral and plant material, with a regular net at one end, for catching drift material; pupae live in hemi-ellipsoidal mineral cases fastened to the substrate, with a silken cocoon firmly adhering to the interior walls of the case.

Most of the species belonging to the Levantine fauna fall into two essentially western Palaearctic species-groups: group *instabilis* and group *guttata*. Two species (*H. pellucidula* and *H. theodoriana*) do not belong to either of these groups. Although not

129

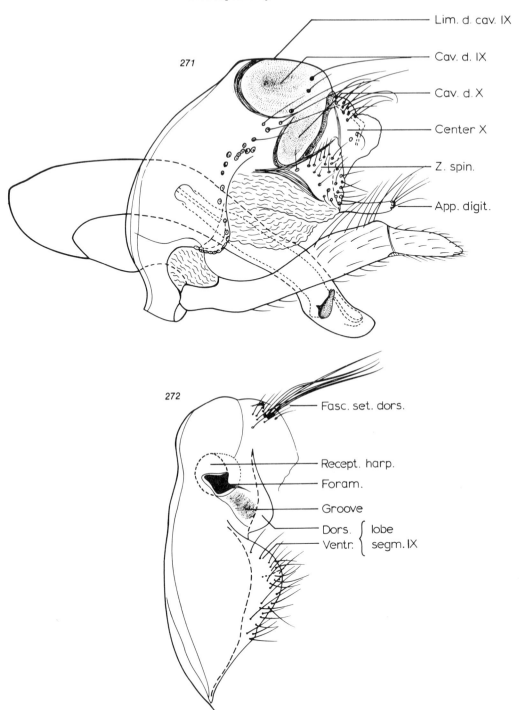

Figs. 271–272: male genitalia of a *Hydropsyche*, and segment IX of female, both lateral view, with terminology used in key and descriptions

very rich, the Levantine fauna is interesting, several species being endemic. A few more species could be present in the province, especially in northern Lebanon and in Syria (see below).

Determining *Hydropsyche* is not an easy task, and this is particularly true for the two species-groups mentioned above, the main factors involved being: a simple morphology of the genitalia, and the fact that non-genital characters are of little use; sometimes an important variability, geographic or not, which can be very interesting but does not make things easier; artifacts determined by the state of preservation of the specimens, or by the more or less strong maceration of the abdomina in KOH (an indispensable operation for this genus!).

On some more species present (or possibly present) in the Levant

A. *H. cornuta* Martynov, 1909. In Malicky & Sipahiler (1984: 210) this species known from the Caucasus and many parts of Asia Minor, was mentioned, without details, from Syria. It is for this reason that it is included here. The male of this species will be easily recognized (central body of segment X with pair of pointed apico-dorsal projections; phallus strongly inflated apically; very short harpago; see Figs. 273–274).

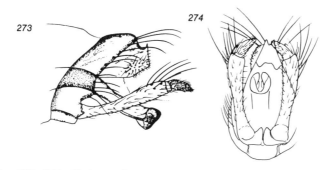

Figs. 273–274: *Hydropsyche cornuta* Martynov, 1909, male genitalia
(273. lateral view; 274. ventral view).
(From Martynov, 1909)

B. *H. kinzelbachi* Malicky, 1980. This species was described (Malicky, 1980a) from the Euphrates (Syria), later mentioned from Iran, too (O'Connor & Dowling, 1968) and possibly does not belong to the Levantine fauna, although it was mentioned from the "Levant" in Malicky (1983: 121). It is, nevertheless, included here. The male may be easily recognized (central body of segment X very strongly projecting apically; strongly sinuous margins of the phallus in dorso-ventral view; very short harpago; see Figs. 275–278).

C. *H. modesta* Navas, 1925 (syn. *H. dissimulata* Kumanski & Botosaneanu, 1974). This species, widely distributed especially in southern Europe, was mentioned from the "Levant", without details, in Malicky & Sipahiler (1984: 210). I lack knowledge of the presence of this species in the Levantine Province, but, of course, its occurrence here is possible (see some comments about it under *H. janstockiana*). *H. modesta* can

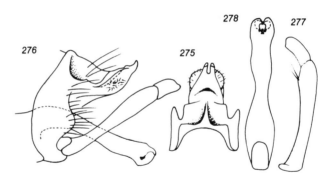

Figs. 275–278: *Hydropsyche kinzelbachi* Malicky, 1980, male genitalia
(275. dorsal view; 276. lateral view; 277. gonopod; 278. phallic apparatus).
(From Malicky, 1980a)

be recognized (male) owing to: dorsal limit of segment IX in lateral view strongly ascendent posteriad and projecting above the deep excision of segments IX + X; phallus very slender, practically straight in dorso-ventral view (see Figs. 279–280).

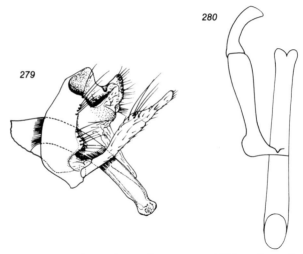

Figs. 279–280: *Hydropsyche modesta* Navas, 1925, male genitalia
(279. lateral view — from K. Kumanski & L. Botosaeanu, 1974,
Acta Mus. Maced. Sci. Nat., 14(2);
280. ventral view — specimen from River Tevere, Italy)

D. I have had the opportunity to examine 3 male *Hydropsyche* caught in the middle course of the Orontes River, in Syria (near Qssair, 1.IV.1979; 1 male, wing expanse 22.5 mm; genitalia: Figs. 281–282; and east of Ain Slimo, 22–25.III.1979; 2 males, wing expanse 18 and 19.2 mm; genitalia: Figs. 283–284 and 285–286; all leg. R. Kinzelbach; sent by H. Malicky). They certainly represent a taxon belonging to what seems to be the superspecies here mentioned under *H. janstockiana*. I have little doubt

Hydropsychidae: Hydropsyche

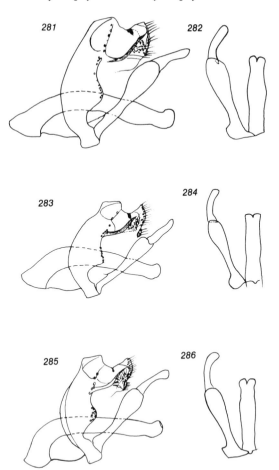

Figs. 281–286: male genitalia
(lateral view, ventral view, figures simplified)
of 3 different specimens of *Hydropsyche* sp. from River Orontes, Syria
(281–282: specimen from near Qssair;
283–284 and 285–286: specimens from Ain Slimo)

that the 3 specimens belong to a single taxon. There is, nevertheless, a difference between one specimen (Figs. 285–286) with a long harpago measuring 1/2 of the length of the coxopodite, and the two remaining specimens with a distinctly shorter harpago: this points to an important variability. I was tempted to consider this taxon as being *H. gracilis* (Martynov, 1909)* but subsequently abandoned this idea, and I now consider the Orontes River specimens as being *H. janstockiana* (Dr W. Mey, Berlin, reached a similar conclusion on examining my drawings and making use of his personal experience). I am convinced that *H. modesta* Navas is distinct from *H. janstockiana* Botosaneanu, although closely related to it: the former is also geographically segregated from the latter species. Anyway, I feel unable to reach a definite decision on the population from the Orontes River solely on the basis of the study of three specimens.

---

* *H. gracilis* Martynov, 1909, originally described from the Caucasus as a subspecies of *H. ornatula* (Martynov, 1909: 540–541, figs. 46–48) is preoccupied by a species of Banks, 1899 (now in *Cheumatopsyche*). Malicky (1977: 7–8, Pl. 6; 1979: 11) renamed it *H. sciligra*, but at the same time — for reasons unknown to me — described it as a new species based on a holotype from Iran. This species (if really only one species is involved) is poorly known, variable, mentioned from the Caucasus and Transcaucasia, parts of Central Asia, Afghanistan, northern Iran, and Asia Minor. In a recent synthesis on the caddisfly fauna of the Caucasus (Kornouhova, 1986: 63, 75), the name *H. gracilis* is retained for the Caucasian species.

Key to the Species of Hydropsyche in the Levant

♂♂

1. IXth dorsum forming a large, rounded plate, without median keel; phallus with characteristic lateral shape (very sinuous, with strongly rounded apex) and exceedingly swollen anteapically in ventral view; a small species.   **H. theodoriana** Botosaneanu
— These characters not present   2
2. From segment X a pair of digitiform appendages extending posteriad 3 (gr. *instabilis*)
— No digitiform appendages extending posteriad from segment X   4
3. (a). "Cav. d. X" deeper than "cav. d. IX"; apical part of phallus — lateral view — narrowing to subacute apex; ratio apical part/entire length of phallus = ca. 1:4.4; moderately pale species.   **H. instabilis** (Curtis)

    (b). "Cav. d. X" shallower than "cav. d. IX"; phallus with slightly obliquely truncate apex; ratio apical part/entire length of phallus = ca. 1:3.7; rather pale species.

    H. jordanensis Tjeder

    (c). "Cav. d. IX" and "cav. d. X" both deep, but that of segment IX distinctly larger; phallus — lateral view — strongly bent downwards preapically, apical part resembling a wooden shoe; a dark species.   **H. longindex** Botosaneanu & Moubayed
4. Phallus with large lateral projections at base of its apical part; in ventral view, apical part of phallus narrowing distad and flattened dorso-ventrally; "cav. d. IX and X" forming together a large, rounded hollow; a large species.   **H. pellucidula** (Curtis)
— Phallus anteapically (ventral view) only moderately swollen; in ventral view, apical part of phallus capitate; "cav. d. IX" and "cav. d. X" not forming together a large, rounded hollow; smaller species   5
5. In lateral view, phallus bent in the normal manner for many *Hydropsyche*: first dorsad, then ventrad; ratio length of harpago/coxopodite = 1:2.

    H. janstockiana Botosaneanu
— In lateral view, phallus bent in a very peculiar manner: first ventrad, then dorsad; ratio length of harpago/coxopodite= 1:2.6.   **H. batavorum** Botosaneanu

♀♀

(Exclusively characters of segment IX; female of *H. batavorum* unknown; key to be used with caution; careful comparison with Figs. 303–305 is compulsory when species in the *instabilis*-group are involved)

1. "Recept. harp." and its "foramen" elongated, distinctly longitudinal; "foramen" reaching directly — i.e. without a "groove" — dorsal end of dorsal lobe of segment IX.

    H. theodoriana Botosaneanu
— These characters not present   2
2. "Recept. harp." small, vertical, cylindrical with rounded bottom, "foramen" distinctly hemi-ellipsoidal; "groove" very distinct, from distal half of foramen to space between the two lobes of segment IX.   **H. janstockiana** Botosaneanu
— These characters not present ("recept. harp." always globose, "foramen" not hemi-ellipsoidal)   3
3. Dorsal lobe IX high (i.e., its length measured on the vertical line) but short, not protruding posteriorly, distal margin quite evenly rounded; ventral lobe IX with central part distinctly protruding; large species   **H. pellucidula** (Curtis)

– Dorsal lobe IX of variable size, but always protruding posteriorly; ventral lobe IX without central part distinctly more strongly protruding posteriorly   4 (gr. *instabilis*)

4. "Fasc. set. dors." composed of fine, shorter setae; " foramen" oblique, elongated; "groove" narrow, from ventral end of foramen to space between the two lobes of segment IX.   **H. instabilis** (Curtis)

– "Fasc. set. dors." also with some long, coarse setae; "foramen" roughly rhombic, proximally with distinct "point"; "groove" broad, from distal limit of foramen to dorsal lobe IX   5

5. Dorsal lobe IX extremely small compared to the enormous ventral lobe; "groove" with parallel margins; a paler species.   **H. jordanensis** Tjeder

– Dorsal lobe IX distinctly smaller than ventral lobe, but difference much less notable than in preceding species; "groove" broadening towards its distal end; a darker species.
                                                                              H. longindex Botosaneanu & Moubayed

Hydropsyche instabilis (Curtis, 1834)
Figs. 287–290, 303

Philopotamus instabilis Curtis, 1834, *Phil. Mag.*, 4:213.
Hydropsyche instabilis —. Curtis, 1836, *Br. Ent.*, 13, text to Pl. 601.

Wing expanse in specimens from Lebanon: ♂ 19.5–23.5 mm (exceptionally larger), ♀ 25–29 mm. The ♂♂ from Naḥal 'Arugot are considerably smaller: 17.5–19 mm. Wings and body, including genitalia, moderately pale.

Male genitalia: The combination of three characters is really important for enabling distinction from the two other Levantine species belonging to the same group, and especially from the very similar *jordanensis*: a) dorsal depression of segment X ("cav. d. X") very well delimited all around, as large as but even deeper than that of segment IX; b) apical part of phallus (lateral view) clearly narrowing to a subacute apex very slightly turned upwards; c) apical part of phallus — measured from the line basally uniting its strong lateral projections — relatively shorter in ventral view (ratio apical part/entire length of phallus about 1:4.4).

Female genitalia (segment IX): Setae in the dorso-apical tufts ("fasc. set. dors.") fine, moderately long. "Recept. harp." slightly oval. Foramen distinctly elongated, oblique, ventral half of its distal (dorsal) limit sinuous. Groove from ventral end of the foramen to dorsal end of ventral lobe of segment IX, short and relatively narrow. Ventral lobe of segment IX much higher than dorsal lobe, but this latter lobe rather well protruding.

Remarks: This species, like the two following ones, belongs to the group bearing its name. It is more similar to *H. jordanensis* Tjeder than to *H. longindex* Botosaneanu & Moubayed. Despite its vast distribution, the species is scarcely variable, at least in the structure of its male genitalia. An exception is represented by the specimens from Naḥal 'Arugot and Naḥal Dawid, with several interesting genital peculiarities (Figs. 589–590).

Distribution: Known from most of Europe, excepting its northern zones; mentioned also from Asia Minor and northern Iran, some of these mentions doubtful. In the Levant, it was caught in numerous Lebanese localities, from almost sea level to ca. 1,200 m a.s.l.; the known localities are in the catchments of Nahr ed Damour (main

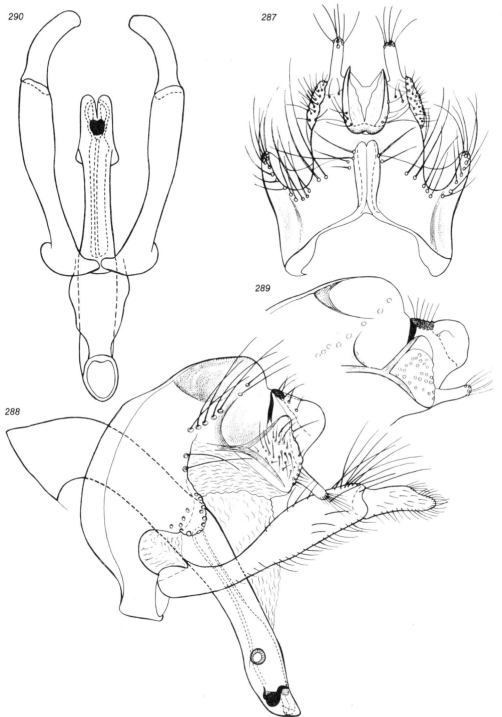

Figs. 287–290: *Hydropsyche instabilis* (Curtis, 1834), male genitalia
(287. dorsal view; 288–289. lateral view, and detail from another specimen; 290. ventral view)

stream and tributaries), of Nahr el Aouali (idem), of the Litani (the Litani itself at Sir el Gharbii(?), and its tributary Yahfoufa), and of the Orontes (main stream and small tributaries). Only two localities are known from Israel: one male specimen was caught by R. Ortal in Naḥal Dan (1); larger numbers were caught in Naḥal 'Arugot (13), a desert stream on the western shore of the Dead Sea, at ca. 350 m below sea level, perhaps the southernmost point in the species' distribution. *H. instabilis* will be found in other parts of the province; however, it is almost certainly only very sparsely represented in the Jordan River catchment (replaced by *H.jordanensis*).

Ecological Notes: A eurytopic species, present from the Crenal to the Epipotamal, from desert streams to moderately cool springs, brooks, and streams in the Lebanese mountains. It seems to tolerate a certain amount of organic pollution. Large populations are commonly developed. The adults are attracted by artificial light; they were caught between March and the first days of December.

Hydropsyche jordanensis Tjeder, 1946
Figs. 270, 291–298, 304

Hydropsyche jordanensis Tjeder, 1946a, *Ent. Tidskr.*, 1964:154–156. Tjeder, 1946b, *Opusc. Ent.*, 11:134–135; Botosaneanu & Marinkovic-Gospodnetic, 1966, *Annls Limnol.*, 2(3):Fig. 11. Type Locality: "... on the River Jordan where ... flowing out the Sea of Galilee, 200 m under the sea level". Type deposited in the Beth-Gordon Museum, Deganya A, Israel.

Wing expanse: ♂ 16.7–20.5 mm, ♀ 21.2–25 mm. Habitus: Fig. 270. Wings and body, including genitalia, moderately pale; forewings very setose, strongly mottled.

Male genitalia: The combination of three characters is really important for enabling distinction from the two other Levantine species belonging to the same group, and especially from the very similar *H. instabilis*: a) dorsal depression of segment X ("cav. d. X") as large as that of segment IX but less well delimited and shallower than the latter; b) apical part of phallus in lateral view only very slightly narrowing to an apex which is slightly obliquely truncate; c) apical part of phallus — measured from the line basally uniting its strong lateral projections — relatively longer in ventral view (ratio apical part/entire length about 1:3.7).

Female genitalia (segment IX): Setae in the dorso-apical tufts ("fasc. set. dors.") very long. "Recept. harp." slightly oval. Foramen roughly rhombic, proximally with a distinct "point", distal (dorsal) limit deeply sinuous. Groove from the distal (dorsal) limit of foramen to dorsal lobe of segment IX not long but broad, margins parallel. Dorsal lobe of segment IX very small compared to the ventral lobe (it may, nevertheless, be somewhat more protruding than represented in Fig. 304).

Remarks: This species, like the preceeding and following ones, belongs to the *instabilis* group. It is more similar to *H. instabilis* (Curtis) than to *H. longindex* Botosaneanu & Moubayed. Some differences may be observed in the male genitalia of various specimens, but the apparently "different aspects" are mostly artifacts. However, Figs.

138

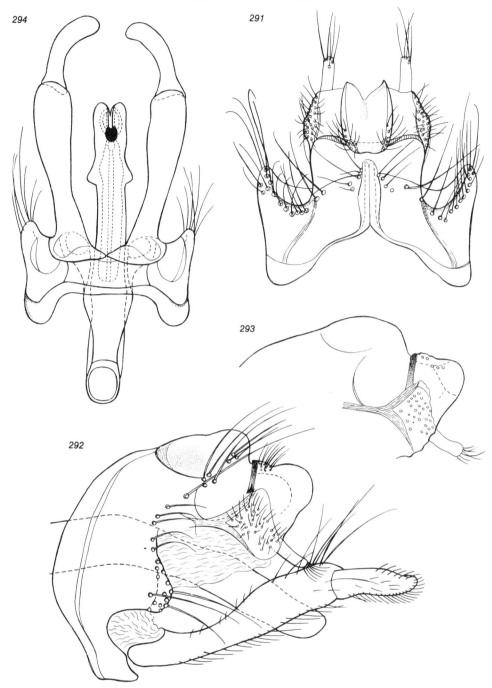

Figs. 291–294: *Hydropsyche jordanensis* Tjeder, 1946, male genitalia,
specimens from Ḥazbani, Israel
(291. dorsal view; 292-293. lateral view and detail from another specimen;
294. ventral view)

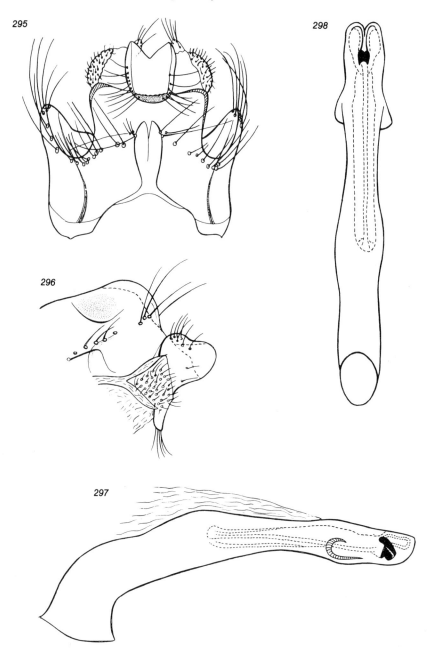

Figs. 295–298: *Hydropsyche jordanensis* Tjeder, 1946, male genitalia,
slightly aberrant specimen from Nahal Meshushim, Israel
(295. dorsal view; 296. lateral view, distal parts only;
297. phallic apparatus, lateral view; 298. phallic apparatus, ventral view)

295–298 depict a slightly aberrant specimen, from Naḥal Meshushim, whose phallus (Figs. 297–298) has, nevertheless, the most typical shape for *H. jordanensis*; it was caught together with typical specimens.

Distribution and Ecological Notes: This species is restricted to the Jordan River catchment — an interesting biogeographic fact (it was erroneously mentioned for Anatolia, by Malicky & Sipahiler, 1984, and by Sipahiler, 1987; fig. 4 in this last publication clearly shows that another species is involved). Almost all the available specimens were caught along streams in the upper and middle reaches of the Jordan, at moderate altitudes. In area 1, the species inhabits, in large numbers, the three main water courses forming the Jordan: [Naḥal Senir (Ḥazbani), Naḥal Dan and Naḥal Ḥermon (Banias)], and it was also caught at Bet Hillel, in the Ḥula Valley, as well as on Naḥal Beẓet (Wadi Karkara). Wadi Faria (6) is an important right side tributary of the Jordan in its middle reach. Several localities are known from area 7, for instance the complex of small streams at Naḥal Meshushim at the foot of the Golan Heights, Buteicha, Naḥal Tavor and an aqueduct near Bet She'an (all with considerable populations). The species was also caught at Senir (Golan Heights, 18), and at Naḥal Tavor, a tributary of the Jordan in its lower course. An interesting question — and one which, unfortunately, I am unable to answer satisfactorily here — is the following: does *H. jordanensis* inhabit the Jordan River itself? The specimens caught by Y. Palmoni during 1935–1938 almost certainly emerged from the Jordan just below Lake Kinneret. There are also a few specimens collected later, during the 1960s, and labelled "Deganya B" or "Jordan River at the Deganya A dam", but I am not quite certain that they emerged from the main water course. In 1984 I tried to attract *H. jordanensis* with an UV lamp at one of the very few points where the river still has a rather fast current, at Bet Zera' bridge, some 2.5 km below the Kinneret: the only *Hydropsyche* caught was *H. janstockiana*. The two species are sometimes caught together by artificial light, but this reveals nothing about their ecological requirements. I strongly suspect that *H. jordanensis* no longer (or almost no longer) belongs to the fauna of the Jordan River, being presently strictly rhithrobiotic. However, a careful field study has to be carried out in order to find the correct answer to this question. The species is on the wing throughout the whole year.

Hydropsyche longindex Botosaneanu & Moubayed, 1985
Figs. 299–302, 305

"*Hydropsyche instabilis* morpha ß du Liban" —. Dia & Botosaneanu, 1983, *Bull. zool. Mus. Amsterdam*, 9 (14):126.

Hydropsyche longindex Botosaneanu & Moubayed, 1985, in: Moubayed & Botosaneanu, 1985, *Bull. zool. Mus. Amsterdam*, 10(11):67–68, Figs. 12–14. Type Locality: Nabaa Bâter ech Chouf, downstream of Niha village, catchment of Nahr el Aouali, Lebanon. Type deposited in the Zoological Museum, Amsterdam.

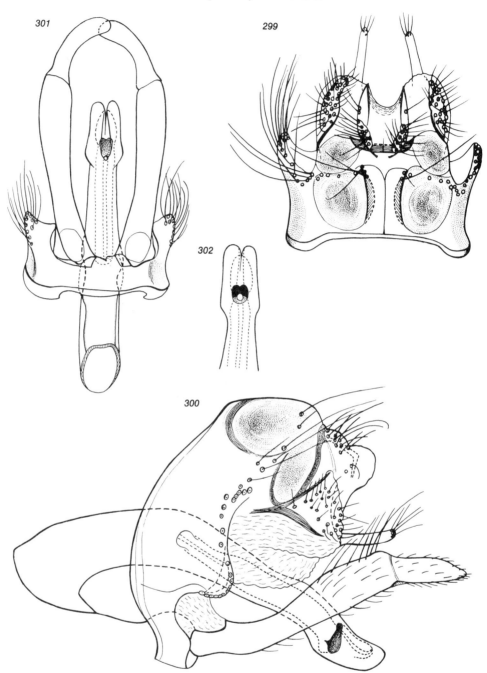

Figs. 299–302: *Hydropsyche longindex* Botosaneanu & Moubayed, 1985,
male genitalia
(299. dorsal view; 300. lateral view; 301. ventral view;
302. apex of phallic apparatus, dorsal view).
(From Moubayed & Botosaneanu, 1985)

Wing expanse: ♂ ca. 22–24.5 mm, ♀ ca. 27 mm. Wings and all parts of body, including genitalia, distinctly darker (brown) than in the two other Levantine species of the same group.

Male genitalia: The combination of the following characters will enable distinction from the two other related Levantine species: (a) dorsal depressions of segments IX and X both very well delimited and deep, but that of segment IX distinctly larger than that of segment X; (b) digitiform appendages ("app. digit.") of segment X particularly long; (c) phallus, in lateral view, with strongly sinuous dorsal margin, strongly bent downwards preapically, apical part shaped like a wooden shoe, or clog; apical part of phallus, in dorso-ventral view, with attenuate lateral projections.

Female genitalia (segment IX): In the dorso-apical setal tufts ("fasc. set. dors.") a bundle of long, strong setae (not only one such seta, as in the original description!) contrasting with a number of short setae. "Recept. harp." rounded. Foramen roughly rhombic, proximally with distinct point, distal (dorsal) limit deeply sinuous. Groove from the ventral end of the foramen to the ventral half of the dorsal lobe of segment IX not long, but broad and broadening towards the ventral (distal) end. Dorsal and ventral lobe of segment IX of similar shape, but the former distinctly smaller than the latter (the difference being, however, much less marked than in *H. jordanensis*).

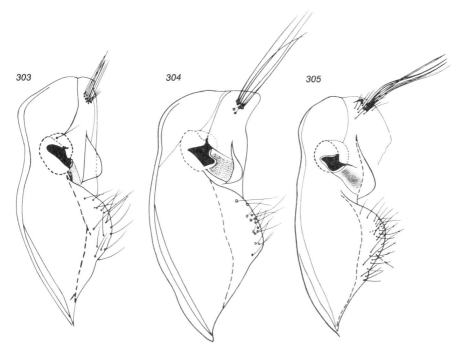

Figs. 303–305: abdominal segment IX of female, lateral view,
in three species of *Hydropsyche* from the *instabilis*-group
(303. *H. instabilis*; 304. *H. jordanensis*; 305. *H. longindex*).
(305: from Moubayed & Botosaneanu, 1985)

143

Remarks: *H. longindex* belongs to the same species group as the two preceeding species. It is very distinct from both in the darker colour and in the male genitalia, whereas the female genitalia are very similar to those of *H. jordanensis*.

Distribution and Ecological Notes: This is probably an endemic Levantine element, and is presently known only from springs and spring-brooks in two small coastal basins of Lebanon: spring and spring-brook Nabaa Bâter ech Chouf, catchment of Nahr el Aouali, 820 m a.s.l.; karst spring Nabaa Mourched, catchment of Nahr el Aouali, 800 m a.s.l.; and Baalechmay, a lateral spring of the stream Beyrouth, 1,000 m a.s.l. We have here the rather exceptional case of an apparently crenobiont *Hydropsyche*.

Hydropsyche pellucidula (Curtis, 1834)
Figs. 306–310

Philopotamus pellucidulus Curtis, 1834, *Phil. Mag.*, 4:213.
Hydropsyche pellucidula —. Curtis, 1836, *Br. Ent.*, Pl. 601 (Text).

Wing expanse (specimens from Naḥal Ḥermon): ♂ 20.5–22.5 mm, ♀ 24.5 mm; these are considerably lower figures than those obtained for example, from specimens from Central Europe: ♂ 22–31 mm, ♀ 27–37 mm (size very variable). Forewings (Fig. 306) in specimens from Naḥal Ḥermon not very setose, membrane uniformly coloured (pale brownish yellow); there is a very important variability in this respect too, a much more distinct pattern being present in Central or Eastern European specimens.

Male genitalia: The following characters will render this species unmistakable: a) dorsal depression of segment IX ("cav. d. IX") and the much smaller upper part of that of segment X together forming a large, deep, clearly delimited, and perfectly rounded hollow (the extended lower part of "cav. d. X" has an irregular shape); b) central body of segment X ("center X") strongly protruding, with dorsal and distal borders regularly rounded; c) the spiny zone of segment X ("Z. spin.") also distinctly protruding posteriorly (distal border of genitalia clearly emarginate). Distal segment (harpago) of gonopods with very blunt apex. The phallic apparatus is characteristically shaped, too, but not easy to describe; it is, in specimens from Israel, only very slightly bent (lateral view), with slightly sinuous dorsal and ventral margins, its distal part strongly tapering to an apex which is very slightly turned upwards; in ventral view, the distal part of the phallus, flattened and slightly narrowing towards the apex, is deeply cleft medially, and at the root of this apical part is located a pair of very strong lateral projections (there is some slight possibility of confusion with certain species of the *instabilis*-group, but *pellucidula* will be immediately distinguished from any species in this group, by the absence of the digitiform appendages of segment X ("app. digit.").

144

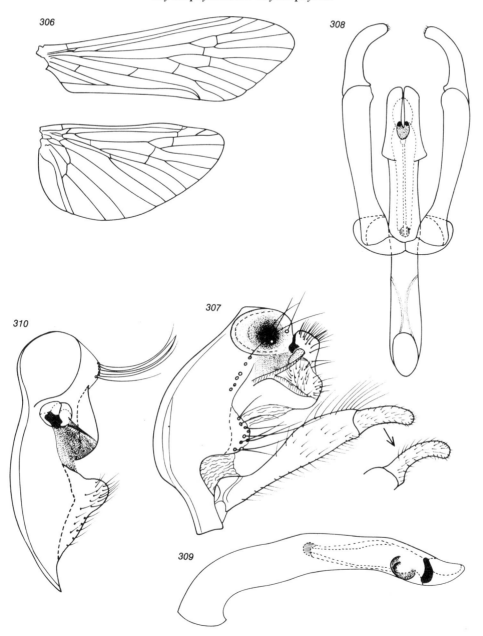

Figs. 306–310: *Hydropsyche pellucidula* (Curtis, 1834),
specimens from Banias, Israel
306. wings; 307–309. male genitalia
(307. lateral view, arrow pointing to a more correct lateral figure of harpago;
308. ventral view; 309. phallic apparatus, lateral view);
310. abdominal segment IX of female, lateral view

145

Female genitalia (segment IX): "Recept. harp." large, roughly globose but irregular, with what seem to be several annex hollows. Its foramen elongated, oblique, all its margins sinuous. Groove distinct though not very deep, broadened distad, from ventral (distal) end of "recept. harp" to lower part of dorsal lobe of segment IX and to space between the two lobes of this segment. Dorsal lobe of segment IX high but only very slightly protruding; ventral lobe slightly higher than it at base, but considerably more strongly protruding, bluntly conical in its central part.

Remarks: *H. pellucidula* belongs to a small but interesting complex of taxa distinguished mainly by the outline of the phallic apparatus, a complex whose existence was discovered only a few years ago, being still very imperfectly studied. It is possible that we are dealing here with a superspecies composed of several, intersterile prospecies; however, in a genus like *Hydropsyche*, they may be valid species. Anyway, the Israeli and Lebanese populations do not belong to one of the several slightly — but clearly — distinct taxa described in recent years from different zones around the Mediterranean. Nevertheless, there are differences between the Levantine specimens examined and, for example, *H. pellucidula* from Central or Eastern Europe; those with regard to size and forewing patterns have already been mentioned; to these, the following may be added: if the male genitalia are, in most of their characters, those of the "typical" *pellucidula*, the very slightly curved phallus differs from that of, say, specimens from Romania, where it is very distinctly bent before its middle (obtuse angle); moreover, it is possible that there are slight differences in the structure of segment IX of the female ("groove" narrower..).

Distribution and Ecological Notes: *H. pellucidula* is mentioned from almost every part of Europe, from the Maghreb and from Asia Minor. But, as already pointed out, in several southern parts of the distribution area of pellucidula (the Dinarids, Corsica, the Iberian Peninsula, Tunisia...) a number of slightly different taxa with apparently restricted distribution occur, mostly replacing the "true" *H. pellucidula*, and one closely related species, first discovered in the Dinarids, seems to have a vast European distribution. The Levantine specimens do not belong to any of these (already described) taxa, but it is not absolutely certain that they belong to the typical *H. pellucidula*.

In the Levant, imagines of this species were caught in Lebanon, more exactly in the middle and lower reaches of the two small coastal rivers Nahr el Aouali and Nahr ed Damour (including one tributary of the Aouali: Nabaa Joun), as well as in the lower course of the Litani, at Sir el Gharbii. The species was also caught in Israel (Upper Galilee, 1), in relatively low numbers; most of the specimens were caught at Nahal Hermon (Banias), together with *H. jordanensis*, but one male was taken, in artificial light, at "Tel Dan" (probably from River Dan itself). Further, several males where caught at Senir (Golan Heights, 18) in the company of *H. jordanensis*. The species certainly prefers water courses belonging to the Metarhithral, the Epipotamal being inhabited, too, if the lower course of the Litani does really belong to this zone. Imagines found on the wing from March to the end of September.

146

Hydropsyche theodoriana Botosaneanu, 1974

Figs. 311–315

Hydropsyche theodoriana Botosaneanu, 1974b, *Israel J. Ent.*, 9:170–173. Type Locality: Hazbani, Israel. Type deposited in the Zoological Museum, Tel Aviv University.

Wing expanse in a larger series of ♂♂ from Nahal Hermon (Hazbani): 13.5–16.4 mm; in the ♀ : ca. 15.5 mm; this is possibly the smallest *Hydropsyche* in the Levant. Slender habitus. A very pale species, wings yellowish, uniformly coloured. Oblique lines only on a small number of antennal segments (generally from 3rd to 7th).

Male genitalia: segment IX characteristically shaped, in dorsal view resembling a "dorsal plate", almost round but narrower at base, with a small indentation in the middle of its posterior border, and with a slightly depressed semicircular posterior zone (the main part of the plate being slightly convex). The 2nd segment (harpago) of the inferior appendages is very slender, medially bent. The shape of the phallic apparatus is also extremely characteristic; in lateral view it is broader at the base, narrowing towards a zone where it is strongly bent downwards, then again strongly broadening — mainly on the ventral side — to the anteapical part which, in dorso-ventral view, is very inflated; the distinctly narrower (ventral view !) phallic apex is clearly rounded (capitate).

Female genitalia (segment IX): Setae in the dorso-apical tufts in a compact bundle, a few of them particularly long. "Recept. harp." very characteristic: elongated, almost longitudinal, reaching directly (i.e., without groove) the dorsal part of the dorsal lobe of segment IX; its foramen elongated, too. Dorsal lobe of segment IX moderately developed, but nevertheless much higher than the ventral lobe, which is conical and well protruding.

Remarks: *H. theodoriana* is a very distinct species; it possibly belongs to a species-group which may merit the name "group *angustipennis*".

Distribution and Ecological Notes: This species is presently one of the most interesting possibly Levantine endemic elements. It was caught in three Lebanese localities belonging to small coastal hydrographic basins: Nahr el Hammam, a tributary of Nahr ed Damour (45 m a.s.l.); Nahr el Aouali, at 230 m a.s.l.; and Ouâdi Râs el Mâ, a stream fed by Nabaa Salman, in the catchment of Nahr el Aouali (800 m a.s.l.). It is, moreover, regularly caught along the Hazbani, one of the headwaters of the Jordan River, in Israel (1), where the population is certainly considerable. In the Hazbani the species coexists with *H. jordanensis*. It is rather strange to note that no specimen has been taken, to this day, from the Dan or the Banias, the other main water courses forming the Jordan River. The species is a typical rhithrobiont of strongly flowing streams, the Hyporhithral being possibly preferred to the Metarhithral. On the wing at least from May to October.

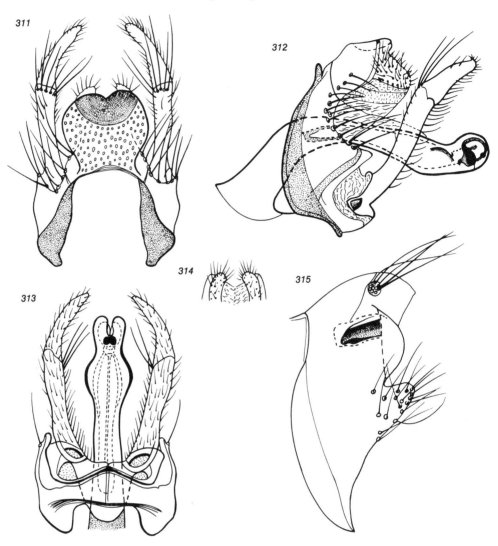

Figs. 311–315: *Hydropsyche theodoriana* Botosaneanu, 1974
311–314. male genitalia
(311. dorsal view; 312. lateral view; 313. ventral view;
314. "z. spin." of segment X, dorsal view);
315: segment IX of female, lateral view.
(311–314: from Botosaneanu, 1974b)

Hydropsyche janstockiana Botosaneanu, 1979
Figs. 316–324

Hydropsyche exocellata auct., nec Dufour, 1841 (e.g., Tjeder, 1946a, *Ent. Tidskr.*:154; Tjeder, 1946b, *Opusc. Ent.*:134; Botosaneanu & Gasith, 1971, *Israel J. Zool.*, 20:102).
Hydropsyche janstockiana Botosaneanu, 1979b, *Bull. zool. Mus. Amsterdam*, 6(21):162–165.
  Type Locality: "Hula Reserve", Israel. Type kept in the Zoological Museum, University of Amsterdam.

Wing expanse: ♂ 17–21.5 mm, ♀ 20–20.6 mm. Membrane of forewings pale (yellowish), relatively slightly setose, setae mostly pale, but in not rubbed specimens a distinct, moderately dark brown reticulation is present.

Male genitalia: The combination of the following characters will enable one to distinguish this species: a) dorsal limit of dorsal depression of segment IX ("lim. d. cav. IX"), in lateral view, always descending posteriad and, at its distal end, never protruding above the (deep, rounded) excision separating segments IX and X; b) long distal segment (harpago) of inferior appendages measuring almost exactly 1/2 of the length of the basal segment; c) phallic apparatus, in either dorso-ventral or lateral view, distinctly inflated anteapically, though not identically in all specimens, with a clear constriction before this swelling, distinctly capitate in lateral view (upper apical angle strongly protruding, rounded; lower apical angle also rounded, but only very slightly protruding). A few additional comments are necessary. Dorso-median keel of segment IX ("car. d. IX") extremely variable (Fig. 317). Dorsal depressions of segments IX and X of similar size and depth (variations in their outline: Figs. 318, 321, 323). Central body of segment X ("center X") in lateral view, protruding posteriad, but only moderately, terminating in a small point. Small spiny zones of segment X ("Z. spin."), and adjacent zones: particularly complex. Harpago slender, turned mediad, with parallel margins, excepting the moderately broadened, obtuse apex. Phallic apparatus with a bend between its 1st and 2nd thirds (lateral view) which is variable in different specimens but generally quite distinct (Figs. 318, 321, 323); in lateral view, the swelling of its preapical part affects the ventral margin more strongly than the dorsal one; lateral margins of distal part, in dorso-ventral view, slightly but visibly diverging.

Female genitalia (segment IX): Setae in the dorso-apical tufts ("fasc. set. dors.") inserted on a larger area, the most ventral ones much shorter than the others. "Recept. harp." small, vertically placed, cylindrical with rounded bottom. Foramen hemi-elliptical. There is an important annex hollow to the "recept. harp.", situated below it and more distad. Groove extremely distinct, long and fairly narrow, from distal half of foramen to space between dorsal and ventral lobes of segment IX (and accompanied proximally by what seems to be a second, less well delimited and shallower groove). Dorsal and ventral lobes of segment IX almost equally high, but dorsal lobe perfectly rounded and more strongly protruding than irregular ventral lobe.

149

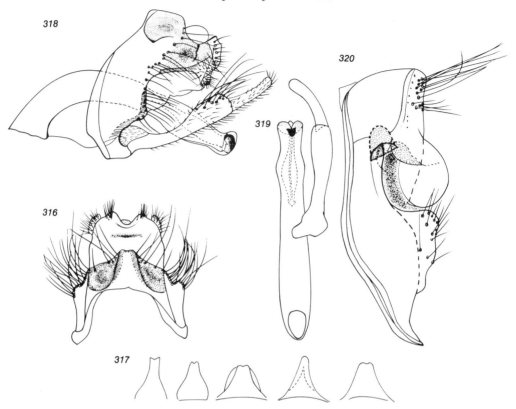

Figs. 316–320: *Hydropsyche janstockiana* Botosaneanu, 1979
316–319. male genitalia
(316. dorsal view; 317. variations in shape of medio-dorsal keel on segment IX;
318. lateral view; 319. ventral view);
320. segment IX of female, lateral view
Figs. 316, 318, 319: holotype from "Hula Reserve"; 317: various paratypes;
320: specimen from Jordan River at Bet Zera' bridge.
(From Botosaneanu, 1979b, except 320)

Remarks: This species belongs to the important and difficult Palaearctic *guttata*-group. Botosaneanu (1979b: 163–164) emphasized the strong similitude to *H. sciligra* Malicky = *H. gracilis* Martynov; other very similar taxa are, in my opinion, *H. demavenda* Malicky, 1977 (described from Iran), and *H. speciophila* Mey, 1981 (described from Central Asia); see also, in the general part on genus *Hydropsyche* in the present book, some comments on certain specimens from the Orontes River, in Syria: these may well belong to *H. janstockiana* interpreted as a superspecies. Dr W. Mey (Berlin) informed me (in lit.) that, in his opinion, *H. janstockiana* is clearly distinct from species like *H. sciligra* (*gracilis*) or *H. speciophila*. I am firmly convinced that *H. janstockiana*, which is almost certainly an endemic element of the Levant, *can* be distinguished from (all?) these taxa by the combination of some stable male genital

150

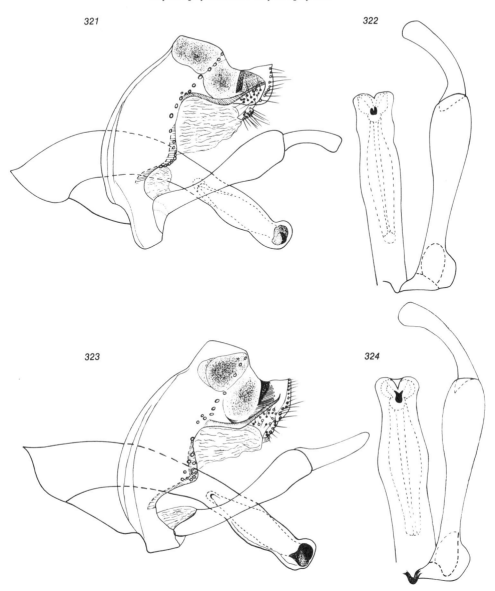

Figs. 321–324: *Hydropsyche janstockiana* Botosaneanu, 1979, male genitalia
(321. & 323. lateral view; 322 & 324. ventral view);
specimen from Jordan River at Bet Zera' bridge (321–322)
and from "near Sea of Galilee, Jordan Valley" (323–324;
this specimen, collected by Y. Palmoni in 1933 and kept in the
"Beth Gordon" Museum, was determined by M.E. Mosely as *H. guttata* Pictet)

151

characters, more or less as shown in the above description. Nevertheless, we are faced here with a very difficult taxonomic problem, and, as already emphasized by Botosaneanu (1979b), we are possibly dealing with a small complex of very closely related taxa inhabiting the considerable territory represented by Central Asia, the Caucasus and Transcaucasia, Iran, Asia Minor, and the Levant. Maybe this is a superspecies composed of potentially or actually intersterile prospecies.

Malicky (1983:122) interprets *H. janstockiana* as subspecies of *H. modesta* Navas, this opinion being apparently based on the presence of hybrids "in the contact zone" (see comments on *H. modesta*, on pp. 131–132). I do not know anything about the existence of hybridization between *H. modesta* and *H. janstockiana*, and definitely cannot believe that these could be subspecies (i.e., geographical races) of a single species. One should not exclude the possibility that *H. modesta*, too, belongs to the superspecies already discussed. However, in my opinion the similarity between these two taxa is less marked than that between *H. janstockiana* and, e.g., *H. sciligra* or *H. speciophila*. One striking character clearly distinguishing *H. modesta* from *H. janstockiana* is its perfectly straight phallus in dorso-ventral view; another is the dorsal limit of the dorsal depression of segment IX, in lateral view, which is markedly *ascending* posteriad. Otherwise, there is much similarity (not identity, however) in the female genitalia.

Distribution and Ecological Notes: *H. janstockiana* is possibly confined to the Jordan River proper. Its apparent absence, as well as that of closely related taxa, from south and central Lebanon is noteworthy; however, see General Remarks on the genus *Hydropsyche*, concerning the specimens from the lower course of the Orontes. The northernmost known locality is H̲uliot (Sedé Neḥemya) in the Upper Galilee (1) at 85 m a.s.l., at the junction of the large streams feeding the Jordan River; further south in the same area, the species was caught, almost at the same altitude, at H̲ula ("Hula Reserve" on the labels, but there is no doubt that the insects, caught in a light-trap, came from the Jordan River). It was also found in the "Jordan Park", slightly north of Lake Kinneret, at Buteicha, as well as at Deganya B, and near Bet Zera' bridge (from 200 m a.s.l. to 225 m below sea level), i.e., in localities very near to the outlet from Lake Kinneret (7). I have also seen one male caught by artificial light at Kibbutz Senir, in the Golan Heights (18). It is, unfortunately, unknown how far downstream the species is present. In the middle and lower course of the Jordan River, *H. janstockiana* is the only representative of the genus; coexistence of this epipotamobiont with the rhithrobiont *H. jordanensis* (see Ecological Notes for this latter species) is apparently an exceptional case, although the two species are sometimes caught together by artificial light. *H. janstockiana*, probably represented by considerable populations in suitable habitats of the river (or in the rare ones still existing!), was found on the wing throughout the year; it is readily attracted by artificial light.

152

Hydropsyche batavorum Botosaneanu, 1979

Figs. 325–327

Hydropsyche batavorum Botosaneanu, 1979b, *Bull. zool. Mus. Amsterdam*, 6(21):162, Fig. 1
    A–D. Type Locality: Deganya A, on Jordan River, Israel. Type specimen preserved in the
    Zoological Museum, University of Amsterdam.

Wing expanse in the unique known ♂ specimen: ca. 21.5 mm. Forewings pale,
membrane beige.
Male genitalia: Dorsal depression of segment IX ("cav. d. IX") small but deep and
well delimited. In lateral view, dorsal limit of this depression slightly ascending
backwards, not protruding above the moderately deep, rounded dorsal excision
between segments IX and X. Central body of segment X (lateral view) with
well-protruding dorso-apical angle. Spiny zones of segment X ("Z. spin.") large. Distal
segment (harpago) of the gonopods bent mediad, short (ratio between its length and
that of the basal segment = 1:2.6). The phallic apparatus is extremely characteristic,
having some unique peculiarities; in dorso-ventral view it is distinctly swollen
anteapically; in lateral view, its outline is distinct from that of all the species of the
guttata-group: in its proximal 2/3 it is sinuous and slightly descending backwards,
without a bend; starting at the level of the anteapical swelling (which affects the
ventral as well as the dorsal margin) it is directed upwards; at the apex, a very distinct,
rounded dorsal projection is not accompanied by a ventral one.
Female unknown.
Remarks: This is a very peculiar species belonging to the *guttata*-group, like *H.
janstockiana*, but clearly distinct from the latter species, as well as from other species
of the same group, in several male genital characters and especially in the phallic
apparatus. Malicky (1983:122) denotes this species with a sign meaning "perhaps a
monstrosity". There is, in my opinion, no evidence supporting this supposition:

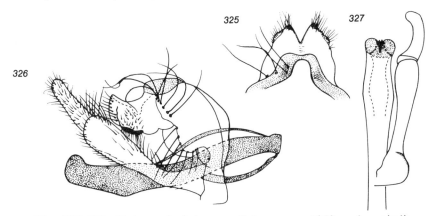

Figs. 325–327: *Hydropsyche batavorum* Botosaneanu, 1979, male genitalia
(325. dorsal view; 326. lateral view; 327. ventral view)
(From Botosaneanu, 1979b)

although the male genitalia of the unique specimen presently known are very slightly damaged, all the parts have a perfectly harmonious, complete, and symmetrical appearance.

Distribution and Ecological Notes: Only one specimen is presently known; it was caught on 7 March 1964, by Y. Palmoni, at Deganya A (Jordan River). We are dealing here either with a remarkably rare species, or with one that has become extinct as a result of the drastic changes in the middle and lower course of the Jordan River in recent decades. Intensive collecting with the aid of artificial light along the river will reveal which one of these two suppositions is correct.

Family POLYCENTROPODIDAE Ulmer, 1903
Abh. Verh. naturw. Ver. Hamburg, 18:117 (as Polycentropinae)

Type Genus: *Polycentropus* Curtis, 1835.
Small or medium sized, stout and very setose insects. Ocelli absent. Maxillary palpi in both sexes long, with short and subequal first two segments, 3rd segment longer, not apically but anteapically inserted on 2nd segment whose apex has stiff setae, 5th segment long, annulate and flexible. Antennae strong, with short segments, scapus scarcely stronger than following segments; they are at most as long as forewings. Spurs setose: 3, 4, 4 (or 2, 4, 4). Tibiae and tarsi of intermediate legs in female more or less dilated. Wings more or less broad, elliptical at apex, both pairs relatively similar, very setose, mostly with a pattern of spots of golden setae; venation with many apical forks, but much variability in this respect; in forewing R1 not forked apically, discoidal, median, and thyridial cells closed, thyridial cell mostly in contact with median cell (hindwings variable in this respect). Abdominal sternite V with lateral filiform appendages (perhaps not always). Male genitalia robust, rather complex; gonopods unisegmented but of more or less complex structure. Female genitalia very short (no ovipositor).

Key to the Genera of Polycentropodidae in the Levant

1.	Forewing with f5, and hindwing with f1.	**Polycentropus** Curtis
–	Forewing without f5, and hindwing without f1.	**Pseudoneureclipsis** Ulmer

Genus POLYCENTROPUS Curtis, 1835
Br. Ent., 12:Pl. DXLIV (text)

Type Species: *Polycentropus irroratus* Curtis, 1835, by original designation.
Diagnosis (mainly based on Palaearctic species): Small or medium sized, stout and very setose insects (Fig. 328), with anterior wings brown, generally irrorated with

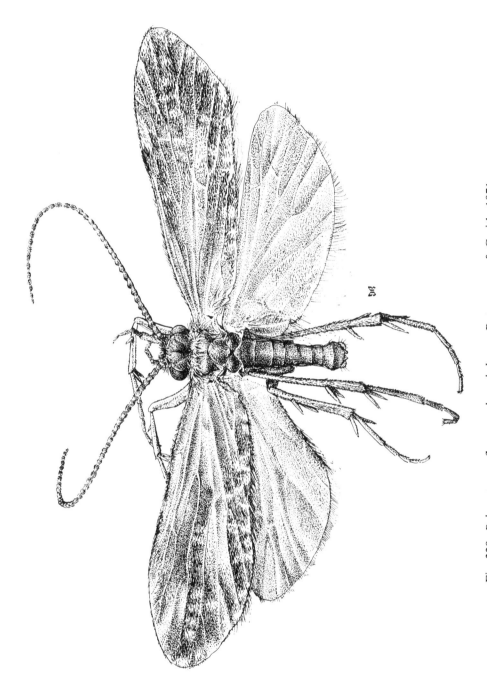

Fig. 328: *Polycentropus flavomaculatus hebraeus* Botosaneanu & Gasith, 1971

distinct golden-yellow round spots. No ocelli. Maxillary palpus with first two segments very short. Antennae thick, distinctly annulated. Spurs 3, 4, 4; intermediate tibiae and tarsi of female greatly dilated. In the forewing: forks 1–5 present; there are transverse veins between C and Sc (two), Sc and R1, and R1 and R2. In the hindwings: forks 1, 2, 5 present; discoidal cell open; first two anal veins coming very close together in their middle, and united there by a transversal vein. Male genitalia variously shaped (to give a general diagnosis is not an easy task): segment IX well developed laterally and ventrally, but not dorsally; segment X entirely membranous; superior appendages small and simple, free; from beneath segment X, a pair of intermediate appendages (sometimes described as "paraproctal sclerites") always present, well developed, sclerotized, often strongly curved downwards at their end; inferior appendages of variable size, always bilobed in some way, hollow, with complex structure especially if viewed from behind; phallic apparatus of very varied structure. Female genitalia very short (stout); the ventral lobes, or plates, of segment VIII are far apart, not connected; segment X small, with three pairs of digitiform appendages.

General Remarks. *Polycentropus* is a relatively large genus, present in many parts of the world, but especially well represented in the Holarctic (with some 30 western Palaearctic species). This is a heterogeneous genus, in need of a revision, and almost nothing is known of the relationships between its species, American species excepted; the two known Levantine species belong to distinct species-groups. The larvae build loose, sometimes conspicuous, funnel-shaped retreats and nets in lotic (sometimes lenitic) habitats, designed for catching the drift; the pupae live in cocoons firmly adhering to the inside of hemi-ellipsoidal pupal cases built from mineral fragments. The adults are very active insects.

Two other genera are closely related to *Polycentropus*: *Plectrocnemia* Stephens, 1836, and *Holocentropus* McLachlan, 1878. Both are considered as *Polycentropus* by American authors, although they certainly merit at least the status of subgenera, representing distinct natural lineages. Since both these genera may well be found in the Levant, especially in its northernmost parts, the most important non-genital distinctive characters will be mentioned here: *Plectrocnemia* will be recognized by the closed discoidal cell in posterior wings, and by the scarcely dilated intermediate tibiae and tarsi in the female; *Holocentropus* by the absence of f 1 in posterior wings; in both these genera, the first two anal veins in the hindwings are not very close together in their middle, and not united there by a cross-vein.

Key to the Species of Polycentropus in the Levant

♂♂

1. Sternite IX deeply excised proximally; superior appendage roughly rounded, without appended sclerotized hook; inferior appendages, in lateral view, distinctly longer than remaining parts of the genitalia.   **P. flavomaculatus hebraeus** Botosaneanu & Gasith
    Sternite IX not excised proximally; superior appendage roughly triangular, with a sclerotized hook attached to it; inferior appendages, in lateral view, not protruding beyond remaining parts of the genitalia.   **P. baroukus** Botosaneanu & Dia

156

♀♀

1. Ventral lobes (or plates) of segment VIII, in ventral view, ovoidal, without emarginated
   lateral border.　　　　　　　　**P. flavomaculatus hebraeus** Botosaneanu & Gasith
2. Ventral lobes (plates) of segment VIII, in ventral view, with lateral border emarginated,
   resulting in a more slender distal part of the lobes.　　**P. baroukus** Botosaneanu & Dia

Polycentropus flavomaculatus hebraeus Botosaneanu & Gasith, 1971

Figs. 328–336

Polycentropus hebraeus Botosaneanu & Gasith, 1971, *Israel J. Zool.*, 20:103–106. Type
　　Locality: Tel el Kadi (Tel Dan), Israel. Type deposited in the Zoological Museum, Tel Aviv
　　University.
Polycentropus flavomaculatus hebraeus —. Botosaneanu & Malicky, 1978:344.

Wing expanse: ♂ about 13–13.5 mm, ♀ about 14.5 mm. Habitus: Fig. 328. Wings:
Fig. 329. Forewings covered by reddish-brown setae, with very numerous small,
round, golden-yellow spots, uniformly distributed and in contrast with the
background colour. On the sides of abdominal segment V, in both sexes, a pair of long
filaments (Fig. 330) — not to be confused with the branchial rudiments found on the
abdominal segments of the adult insects.

Male genitalia. Sclerotized part of segment IX, in lateral view, with convex proximal
margin; sternite IX is proximally very deeply excised in its middle. Triangular in
dorsal view, the superior appendages are roughly round in lateral view; they have
proximally a strong "ridge" whose distal end forms a well protruding dorso-proximal
lobe of the appendage. Median lobe of segment X membranous, distally trilobed.
Intermediate appendages ("paraproctal processes") in dorsal view slightly convergent
posteriad; their apical parts are robust, in lateral view strongly bent at acute angles,
tips pointing downwards and laterad (with only a slight anteriad tendency); just before
the tips, a minute wart with minute spines (a strong seta inserted on the dorsal margin
near the bending point). Inferior appendages long, much longer than superior
appendages, deeply excised apically, bottom of the sinus oblique, superior lobe only
half the length of, and more slender than, inferior lobe; in ventral view the inferior
lobes are roughly rectangular (the structure of these appendages is better seen in Fig.
334). Phallic apparatus: Fig. 335.

Female genitalia typical for a *Polycentropus*. "Ventral plates" of segment VIII roughly
ovoidal, without excision of lateral margin, and therefore without more slender distal
part.

Remarks: This was originally described as independent species, but it was
subsequently decided that it merits only the status of a subspecies of *P. flavomaculatus*
(Pictet, 1834). Because of the complex nomenclatorial problems with this species, the
following synonymy is perhaps useful (we make an exception by publishing it here):

157

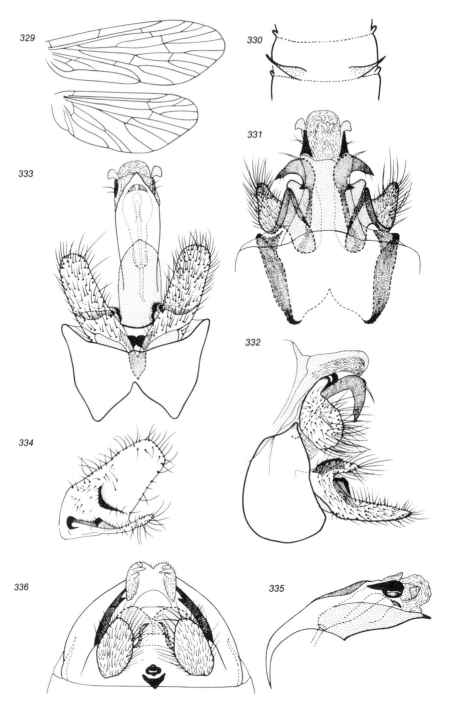

Figs. 329–336: *Polycentropus flavomaculatus hebraeus* Botosaneanu & Gasith, 1971
329. wings; 330. lateral filaments on abdominal sternite V;
331–335. male genitalia (331. dorsal view; 332. lateral view; 333. ventral view;
334. dorsal and median face of left gonopod; 335. phallic apparatus, lateral view);
336. female genitalia, ventral view.

(330: from Gasith, 1969; other figures: from Botosaneanu & Gasith, 1971)

Hydropsyche flavomaculata Pictet, 1834, *Rech. Phryg.*, p. 220, Pl. 19, fig. 2.
Philopotamus multiguttatus Curtis, 1836, *Br. Ent.*, 12:Pl. 544 (nec McLachlan).
Polycentropus flavo-maculatus —. McLachlan, 1878, *A Monographic Revision & Synopsis*, pp. 398–399, Pl. XLII, fig. 1–8.
Polycentropus flavomaculatus —. Neboiss, 1963, *Beitr. Ent.*, 13(5/6):612–613.

P. f. hebraeus can be distinguished from the nominate subspecies by several slight but apparently constant characters of the male genitalia: superior appendages with distal margin rounded (slightly depressed in the nominate ssp.); intermediate appendages stout in dorsal and in lateral view (more slender in the nominate ssp.); inferior lobe of gonopod pointed at apex in lateral view (shorter, stouter, with truncated apex, in the nominate ssp.).

Distribution: *P. flavomaculatus flavomaculatus* is very widely distributed in Europe and present in North Africa too. *P. f. hebraeus* is known — with some degree of certainty — from some Aegean islands (Crete, Naxos ...), Asia Minor, and the Levant. The boundary separating it from the nominate subspecies is very unsatisfactorily known: it must be somewhere in the Aegean Sea, or in southern Greece. According to Malicky (1983:78), *P. auriculatus* Martynov, 1926, from the Caucasus, is also a subspecies of *P. flavomaculatus*.

In the Levant, *P. f. hebraeus* is known from two localities in the Bekaa Valley (Lebanon), as well as from one locality in the upper course of the Jordan River (Israel); it is here probably the sole *Polycentropus*, being replaced in the small coastal basins of Lebanon by *P. baroukus*. The Lebanese localities are: the Orontes at Hermel, downstream from the Zarka spring, at 650 m a.s.l.; and a stream fed by a karst spring at Baalbek (about 1,100 m a.s.l.). The Israeli locality — being probably the southernmost point in the species' distribution — is the important complex of springs and streams at Tel Dan (Tel el Kadi) in Upper Galilee (1), at 200 m a.s.l.

Ecological Notes: An inhabitant of springs, spring-brooks, and larger streams fed by abundant springs, this subspecies certainly needs colder, well-oxygenated, and fairly fast-running watercourses for its development. Downstream from the Zarka spring the Orontes is a river about 10 m broad; the stream at Baalbek is about 2–3 m broad; at Tel Dan, a large complex of springs, spring-brooks, and larger springs is involved, and a large population is certainly present. Adults were caught from February to May, as well as in October. According to Gasith (1969), the species completes one generation in 2–3.5 months.

Polycentropus baroukus Botosaneanu & Dia, 1983
Figs. 337–344

Polycentropus baroukus Botosaneanu & Dia, in: Dia & Botosaneanu, 1983, *Bull. zool. Mus. Amsterdam*, 9(14):131–132. Type Locality: Nabaa Aazibi, spring of Nahr Aaray, basin of Nahr el Aouali, Lebanon. Type deposited in the Zoological Museum, University of Amsterdam.

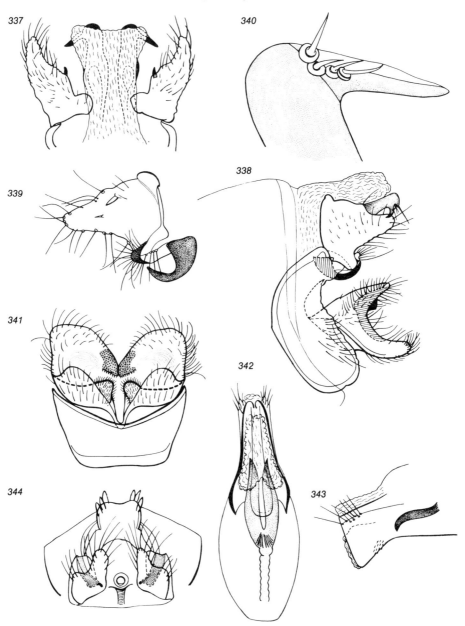

Figs. 337–344: *Polycentropus baroukus* Botosaneanu & Dia, 1983
337–343. male genitalia (337. dorsal view; 338. lateral view;
339. median face of left superior appendage, with its sclerotized hook;
340. apex of intermediate appendage, or "paraproctal sclerite", strongly magnified;
341. segment IX and gonopods, ventral view; 342. phallic apparatus, dorsal view; 343. its apex,
lateral view); 344. female genitalia, ventral view.
(From Dia & Botosaneanu, 1983)

Wing expanse: ♂ ca. 17 mm, ♀ ca. 23 mm.

Male genitalia: Sclerotized part of segment IX, in lateral view, with parallel margins, but with dorsal end distinctly turned posteriad; the sternite is not at all emarginated proximally. Superior appendages roughly triangular, but with various irregularities of their margins; proximal-ventral angle particularly well produced, forming an annex lobe on the median side of the appendages; to this lobe, a characteristic, strong, upward curving sclerotized hook is appended. Median lobe of segment X elongated, entirely membranous, slightly emarginated distally. Intermediate appendages ("paraproctal sclerites"), in dorsal view, two robust, parallel sclerites, apically like rather long, sharp points distinctly directed downwards and obliquely forewards; a group of spines at the bending point. In lateral view, the inferior appendages are very short, shorter than the superior appendages, and they are deeply excised, bottom of the sinus almost vertical, both lobes of similar length, superior lobe more slender than inferior lobe; the complex structure of the inferior appendages is represented in Fig. 341. Phallic apparatus: Figs. 342–343.

Female genitalia typical for a *Polycentropus*. "Ventral plates" of segment VIII with lateral margin somewhat excised, making the distal part of the plates more slender than the proximal one. In the original description, the absence of digitiform appendages on segment X was mentioned, but this was an error (corrected in Fig. 344).

Remarks: *P. baroukus* belongs to a group of species distinct from that containing *P. flavomaculatus hebraeus*, and characterized, inter alia, by the presence of an accessory appendage, generally hook-shaped, rooted at the inferior-median part of the superior appendage. It is presently impossible to offer suggestions about its possible sister-species.

Distribution and Ecological Notes: This species was caught in several localities, but exclusively in the catchments of Nahr el Aouali and Nahr ed Damour, two small coastal rivers of Lebanon. In the basin of Nahr el Aouali the localities are: the river itself upstream of the bridge at Jdaidet ech Chouf; the karst spring Nabaa Mourched; the stream Nabaa Salman near Harêt Jandal ech Chouf; the spring and spring-brook Nabaa Abou Kharma; the spring Nabaa Aazibi of Nahr Aaray. Two localities are known in the basin of Nahr ed Damour: the river itself at Jisr el Qadi, and its left-bank tributary Nahr el Hamman. The species was caught in large numbers, and is almost certainly the sole representative of the genus in these coastal basins. Its ecological spectrum is remarkably wide, ranging from springs to the largest water courses belonging probably to the Hyporhithral; it was caught from ca. 1,000 m a.s.l. almost to sea level. On the wing at least from March to November.

Genus PSEUDONEURECLIPSIS Ulmer, 1913

Notes Leyden Mus., 35:84

Type Species: *Pseudoneureclipsis ramosa* Ulmer, 1913 (by monotypy).

Diagnosis: Small and gracile species, wing expanse rarely exceeding 10–11 mm. Both wings moderately narrow, hindwings shorter than forewings, not broader than them; forewings densely covered with fine, mainly uniformly coloured pubescence, concealing the venation; hindwings paler, more sparsely pubescent. In the forewing furca 5 always absent, f 1–4 generally present (only exceptionally is one of them absent), discoidal, median, and thyridial cells all closed, median cell generally not reaching basally the thyridial cell (but there are several exceptions to this); in the hindwing, costal margin with a very distinct projection in its middle, f 2, 3, 5 present, discoidal and median cells open. Spurs 3, 4, 4; tibia and tarsus of intermediate legs in the female slightly dilated. Male genitalia not always satisfactorily interpreted in the descriptions, with important differences in the various species-groups (e.g., free superior appendages present or absent; segment X with or without a pair of ventral appendages; phallus accompanied or unaccompanied by a pair of slender sclerotized "titillators"). The following is an attempt to provide a general description: segment IX more strongly developed ventrally than dorsally; segment X of complex structure, generally trilobed (but median lobe sometimes further bilobed; and sometimes a supplementary pair of appendages beneath the dorsal complex); laterally to the lateral lobes of segment X, a pair of superior appendages, free or coalescent with them; inferior appendages with a generally stout main body, and having proximally and dorsally an additional appendage, in various degrees free from (or coalescent with) the main body, and ending in a long point directed mediad; phallic apparatus with or without a pair of lateral spiniform appendages, but always with an internal, sclerotized armature of the endotheca, very diverse in the various species. Female genitalia (unknown for most of the species!) not forming an ovipositor, but stout; sternite VIII not split into two lateral lobes or plates; segment IX with strong lateral extensions directed anteriad; segment X relatively large, with 3 pairs of digitiform appendages.

General Remarks: *Pseudoneureclipsis* is an extremely interesting small genus, with slightly more than 20 known species (a figure which will probably not be considerably modified). The general appearance and venation of its species are homogeneous but the male genitalia are heterogeneous, and a revision would certainly lead to interesting results. The sister-group of *Pseudoneureclipsis* is, possibly, the Antillean genus *Antillopsyche* Banks. Four species are Ethiopian (from Ghana, Zaire, Malawi, and Zimbabwe). A heterogeneous group is represented by four western Palaearctic species (Portugal, Morocco, Israel, and Central Anatolia, respectively). There are, moreover, two eastern Palaearctic species in the Ussuri region. Largest is the number of Oriental species, with one in south Iran, two known from the Philippines, Java, and Sumatra, and at least eight from Ceylon (Sri Lanka). The unique Levantine species is probably more closely related to that from southern Iran than to any other species. The restricted, or even very restricted, distribution of each species is striking. Almost nothing is known about their ecology; most species were caught along various types of running water, but at least two species (one of them being *P. palmonii*) are known to inhabit — but not exclusively — the stony and agitated littoral of lakes. The extremely scarce information published on the larvae, shows that these live in sandy tubes fixed to the substrate, modifying these tubes in some way prior to pupation; in my opinion, the following sentence, concerning *P. palmonii*, should possibly be interpreted in this light "The larvae . . . can be distinguished by the fixed tubular case which they construct out of small stones . . . The fully grown larvae seal their tube prior to pupation and pupate inside" (Gasith & Kugler, 1973:59).

Pseudoneureclipsis palmonii Flint, 1967

Figs. 345–349

Pseudoneureclipsis palmonii Flint, 1967, *Ent. News*, 78(3):74–75. Type Locality: Deganya A, Israel. Type deposited in the U.S.N.M.

Wing expanse, ♂ ♀ : 8–10.5 mm. Antennae very distinctly annulate. Forewings narrow, completely covered with very fine, reddish setae, concealing the venation (not "gray-black" as stated in Gasith & Kugler 1973:59); hindwings much paler, with less dense setae, much shorter than forewings but equally broad, strongly produced in the middle of their costal margin. Venation in general lines typical for the genus; forewing with f1–4 (f1 and f3 as long as their stalks; f2 and f4 sessile), discoidal, median, and thyridial cells all closed, thyridial cell separated — but not widely — from median cell; hindwing with f2, 3 and 5, without closed cells; in both wings, some of the transverse veins are very indistinct, and certain of them are probably not represented in Fig. 345. Spurs: 3, 4, 4.

Male genitalia: Segment IX strongly developed ventrally, very narrow laterally, partly indistinct dorsally. Superior appendages large, simple, broad basally, with distinctly produced lower distal angle. Segment X represented by a shorter, setose, median appendage (triangular in dorsal view, apex truncate in lateral view), and by a pair of lateral, more ventrally placed, longer appendages (broadly triangular in lateral view, slightly capitate in dorsal view). Inferior appendage of slightly irregular shape, very broad (high) except its base which is a short, much narrower stalk; most of the median face is cushion-like (swollen) and covered with extremely dense black spines directed proximad; each gonopod has two quite differently shaped basal appendages, one horizontal and dome-shaped, the other one — appressed to the main body of the gonopod but not coalescent with it — vertical, produced in a long point directed mesad. Phallic apparatus much broader in its proximal than in its distal half, internally with a complex armature whose most distinctive part is represented by 4 strong, short, sclerotized spines; there are no external spines laterally accompanying the phallus.

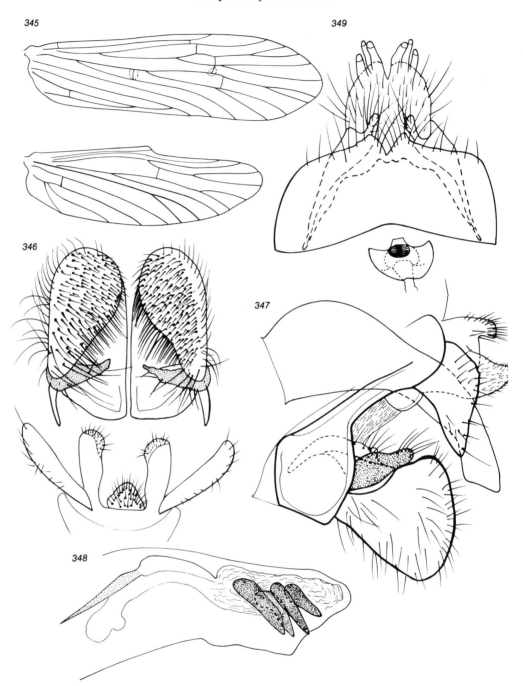

Figs. 345–349: *Pseudoneureclipsis palmonii* Flint, 1967
345. wings; 346–348. male genitalia (346. superior, intermediate, and inferior appendages, dorsal view; 347. lateral view; 348. apical part of phallic apparatus, lateral view); 349. female genitalia, ventral view

Female genitalia not forming some kind of ovipositor. Sternite VIII in the middle of its distal margin with two triangular projections separated by a short cleft. Segment IX well developed only laterally, its two medio-ventral and distal angles produced in digitiform, setose lobes. Segment X relatively large, with 3 pairs of digitiform appendages.

Remarks: *P. palmonii* is related — not very closely — to *P. iranicus* Malicky, 1982, from the province Fars of southern Iran; this is evident from the structure of the male genitalia, but many characters of these genitalia serve to distinguish the two species.

Distribution: The species is one of the most interesting elements of the Levantine fauna, and one of those which are probably true endemics for this province. It is known only from a restricted zone of Israel; most of the known localities are around Lake Kinneret (7), but some other localities appearing either in publications or on labels (Deganya A, Sha'ar HaGolan) are in the immediate vicinity of this lake, and it is almost certain that the insects came from it. Another small group of known localities are: Bet She'an (7; a canal, or aqueduct, with water coming mainly from the spring 'En 'Amal), Naḥal 'Amal, or Assi, in the Bet She'an Valley (5), and the spring 'En 'Amal (Nir Dawid) itself (5). All these localities are between ca. 100 and 200 m below sea level.

Ecological Notes: *P. palmonii* cannot be considered as a strictly limnobiont species, although its most important population(s) may well live in Lake Kinneret; the other localities named above are running water courses of different types, but apparently having the following shared characters: sluggish current, rather warm and eurythermic, more or less turbid water. At Lake Kinneret (Gasith & Kugler, 1973) adults were regularly caught in light traps placed on the shores; larvae and pupae were observed in the littoral "mainly at a depth of 2 m, in many instances almost entirely hidden in the cracks and hollows of the stones". Light trap data from Lake Kinneret led the above-mentioned authors to write that it is "a summer species (June–October) with one peak in September, indicating that in Lake Tiberias *P. palmonii* has only one annual generation". There are, nevertheless, also records from February ('En 'Amal), March (Naḥal 'Amal), May (Bet She'an), and November (Karé Deshe).

Family ECNOMIDAE Ulmer, 1903
Abh. Verh. naturw. Ver. Hamburg, 18:120 (as Ecnominae)

Type Genus: *Ecnomus* McLachlan, 1864.

Small insects with mottled wings (Fig. 350). Maxillary palpi in both sexes with 5 segments, 2nd segment distinctly longer than 1st, 5th very long, annulate and flexible. Antennae moderately strong, at most as long as forewings, with short segments. Spurs: 3, 4, 4 or 2, 4, 4. Intermediate legs of female dilated. Wings moderately narrow, hindwing narrower than forewing; forewing with parabolic apex, R 1 furcate at its end, f 1–5 present in most species, discoidal, median, and thyridial cells closed; in

165

hindwing anterior border without projection or distinct sinuosity, normally only f 2 and 5 present, discoidal, median, and thyridial cells open, space betwen basal parts of SR and M widened. Abdominal sternite V without lateral filamentous appendages. Male genitalia rather complex, gonopods generally simple, but with minor complications. Female without ovipositor, at least in *Ecnomus*.

One genus represented in the Levant: *Ecnomus* McLachlan.

Genus ECNOMUS McLachlan, 1864
Entomologist's mon. Mag., 1:26, 30

Type Species: *Philopotamus tenellus* Rambur, 1842 (by monotypy).

Diagnosis: Small or medium-sized insects (Fig. 350), mostly pale, sometimes with distinct irrorations on the forewings. Maxillary palpus with 1st segment very short, 2nd and 4th longer and similar, 3rd again longer, 5th as long as 2nd, 3rd and 4th together. Wings moderately narrow; forewing with parabolic apex; hindwing narrower than forewing, without a point at its costal margin which is, on the contrary, slightly concave, without dilatation of the anal space. In the venation of the forewings, the most constant distinctive character is the radial vein (R 1) forked at its end; f 1–5 in most species present (rarely f 1 absent); median cell much larger than discoidal cell, in many species basally reaching the thyridial cell (but there are many exceptions to this). In the hindwing, no closed discoidal or median cell, f 2 and 5 normally present, the subradial space (i.e., the space between the basal parts of SR and M) particularly broad. Spurs 3, 4, 4, but in some species 2, 4, 4; on the mid- and hind tibiae internal spurs distinctly longer than external spurs. Midlegs of female dilated. Male genitalia with dorsal and ventral part of segment IX clearly separated; well developed superior and inferior appendages (their relative lengths varying), generally of simple structure but with various minor complications (setose tubercles, spines, small appendages); segment X not strongly protruding distad, moderately complex; phallic apparatus of complex and very diverse structure. Female genitalia never forming an ovipositor, but stout; segment VIII large, segment IX much smaller, both with a pair of apodemes; ventral lobes of segment VIII well developed, medially separated; segment X relatively large, consisting of two distinct halves each with 3 dissimilar digitiform appendages.

General Remarks: *Ecnomus* is a medium-sized genus, very well represented in the Ethiopian and in the Oriental Region (a few tens of species described for each and certainly many more to be described); there are at least 40 species in Australia, but none in the Americas or in New Zealand. The Palaearctic fauna has a very low number of species; if the few known from China, Japan or the Levant are excluded, what remains is one species with wide distribution throughout the western Palaearctic but also penetrating into Japan, India and Ceylon, and 3 species with restricted, peripheral distribution (Iberian Peninsula, northern Africa). The two known Levantine species clearly belong to distinct groups, one of them having obvious

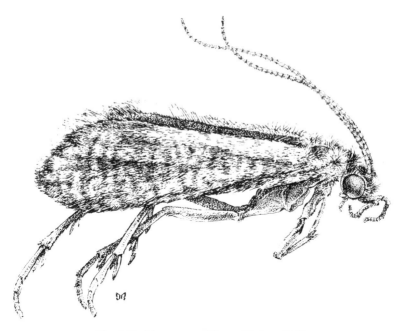

Fig. 350. *Ecnomus galilaeus* Tjeder, 1946

Ethiopian affinities, the other present also in the western parts of the Oriental Region. It is probable that the widely distributed *E. tenellus* Rambur, 1842, will be found in the Levant; one character of its male genitalia allows easy distinction from *E. gedrosicus*: the inferior appendages (lateral view) strongly pointed and curved upwards at their tips. Most of the *Ecnomus* species are typical inhabitants of stagnant or (very) slowly running water. Larvae of *E. galilaeus* and *E. gedrosicus*, according to Gasith and Kugler (1973:58), are spinning irregular nets usually attached to the underside of stones or other suitable substrates; pupae in cocoons sheltered under ovoid, elastic pupal cases built mainly of sand.

Key to the Species of Ecnomus in the Levant

♂♂

1. Spurs 2, 4, 4. Superior appendages roughly rounded; inferior appendages more protruding, with blunt apex. **E. galilaeus** Tjeder
– Spurs 3, 4, 4. Superior appendages elongated, extending to the tip of inferior appendages. **E. gedrosicus** Schmid

♀♀

1. Spurs 2, 4, 4. Ventral lobes of segment VIII very broad at base **E. galilaeus** Tjeder
– Spurs 3, 4, 4. Ventral lobes of segment VIII narrow (pedunculate) at base. **E. gedrosicus** Schmid

167

Ecnomus galilaeus Tjeder, 1946

Figs. 350–359

Ecnomus galilaeus Tjeder, 1946b, *Opusc. Ent.*, 11:135–136. Type Locality: "Eastern bank of River Jordan, near Sea of Galilee". Type deposited in the BMNH.
Ecnomus kunenensis: Botosaneanu, 1963 (nec K.H. Barnard, 1934), *Polskie Pismo ent.*, 33(2):95–98.

Wing expanse, measured from some 50 specimens from Lake Kinneret: ♂ 8.6–10.7 mm, ♀ 10.7–12.4 mm. Habitus: Fig. 350. Colour of practically all parts of the body very pale; this applies also to the forewings, which are uniformly furnished with pale yellowish, not very dense setae. In forewing, the median cell basally reaches the thyridial cell. Spurs: 2, 4, 4.

Male genitalia: In lateral view, a strong distal excision (not distinct in Fig. 353) separates the dorsal from the ventral part of segment IX; sternite IX more strongly protruding distad than the tergite, which does not project above the superior appendages. Superior appendages (also described in *Ecnomus* as "main part of lobes of Xth segment") roughly rounded, inferior margin slightly sinuous; on their median side, distally a large number of strong, black spines, and more proximally numerous setae arising from inflated papillae (as seen in Figs. 352 and 353). Medially to the superior appendages, and above the phallic complex, a pair of appendages ("upper penis cover", or "internal process of lobe of Xth segment") certainly belonging to segment X; they are obliquely directed downwards, somewhat twisted, basally broad, clothed dorsally with numerous small tubercles, and with a digitiform upturned apical part terminating in a few short setae. Inferior appendages longer than superior appendages, broad basally (and there with distinct medio-ventral lobes), then strongly tapering, but apical part in lateral view again broadened, very blunt (in dorsal view slightly beak-like at tip); some strong setae arise from their medio-apical part. Phallic complex in lateral view narrow in its basal, sclerotized parts, very strongly dilated in its distal, membranous parts, with a pair of long, sharply pointing titillators arising near the base and having a distinctly dorsal position.

Female genitalia (described here for the first time): There is no ovipositor. Segment VIII with a pair of well developed apodemes; its ventral lobes have very irregular distal margins furnished with a few strong hairs, and their bases are very long (in other words, each lobe is distinctly more developed proximally than distally, as seen in lateral or ventral view). Segment IX with a pair of long apodemes; in the middle of its distal margin is a pair of large swellings furnished with long setae. Segment X with its two halves very distinct, vertically placed, very setose, with 3 dissimilar digitiform appendages.

Remarks: As pertinently shown by Scott (1968:412–414), *E. galilaeus* belongs to a section of the *natalensis*-group of *Ecnomus*, which includes a number of African species (about 23 described, according to Barnard & Clark, 1986) very similar in their genital and non-genital structures. *E. galilaeus* is certainly very closely related to *E.*

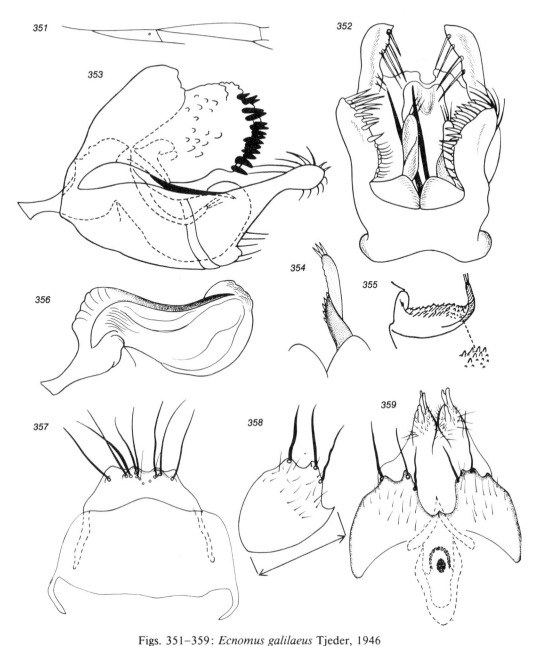

Figs. 351–359: *Ecnomus galilaeus* Tjeder, 1946
351. median and thyridial cells in forewing;
352–356. male genitalia (352. dorsal view; 353. lateral view;
354. "Upper penis cover", dorsal view; 355. the same, lateral view;
356. phallic apparatus, lateral view);
357–359. female genitalia (357. segments VIII and IX, dorsal view;
358. ventral lobe of segment IX, lateral view; 359. ventral view).
(352, 353, 354, 356: inked from originals prepared by Dr. K.M.F. Scott —
Albany Museum, South Africa — from a male paratype,
and kindly made available to the author; 355: from Scott, 1968)

kunenensis K.H. Barnard, 1934 (large distribution, imperfectly known, in south-west and central Africa).

Distribution: *E. galilaeus* is one of the Levantine caddisfly species with obvious Ethiopian origin and affinities; it is probably an endemic element of the Levantine fauna. Exclusively known from northern Israel, being caught from: 'En Te'o and Ḥula (in Upper Galilee: 1); Lake Kinneret at various localities (7); Birket Ram (in the Golan Heights: 18); the Ẓalmon and Netofa reservoirs (in Lower Galilee: 2); and 'En 'Amal, or Nir Dawid (Valley of Yizre'el: 5). The altitudes range from 950 m a.s.l. (Birket Ram) to 210 m below sea level (Kinneret).

Ecological Notes: This is a typical standing-water species: most of the named localities are lakes of various sizes; 'En Te'o, and, probably, 'En 'Amal are large limnocrenous springs, whose large pools with aquatic vegatation could be interpreted as practically stagnant water. I am almost sure that *E. galilaeus* does (and did) not belong to the fauna of the Jordan River, despite the information accompanying the original description. The species was found on the wing from March to November. Gasith & Kugler (1973:59) mention that, according to light trap data, *E. galilaeus* reaches its maximum in April, shows two peaks of population size, and seemingly has two generations a year. The ratio of specimens of the two sexes attracted by artificial light can be quite different from one catch to another.

Ecnomus gedrosicus Schmid, 1959

Figs. 360–366

Ecnomus gedrosicus Schmid, 1959, *Beitr. Ent.*, 9(3/4): Pl. 4, figs. 7–8; 9(5/6):698. Botosaneanu, 1963, *Polskie Pismo ent.*, 33(2):96–97.

Wing expanse, in specimens from Israel: ♂ 11.5–12.8 mm (but one specimen measured 9.6 mm), ♀ 10.7–14 mm. The two ♂ specimens from Iran described by Schmid, have an expanse of only 10.5–11 mm. Practically all body parts are darkish brown, certainly darker than in *E. galilaeus*. Forewings sometimes paler and without distinct irroration, but normally darker and with a distinct pattern of paler, round spots, particularly distinct in dry specimens. Venation: Fig. 360; in the forewing, median and thyridial cells are widely separated. Spurs 3, 4, 4 (proximal spur on foretibia very small).

Male genitalia: Dorsal part of segment IX oblique, protruding above base of superior appendages in lateral view; ventral part more strongly developed, not separated from dorsal part by a distinct apical excision. Superior appendages long (extending to tip of inferior appendages), broader basally than at the very obtuse apex, of simple structure, slightly curved mediad at the tips which are furnished on the median side with strong setae. Segment X is apparently represented by a pair of short, forked appendages ("appendices sus-péniaux" of Schmid) arising medially from below the bases of the superior appendages and obliquely directed downwards (inner branch

170

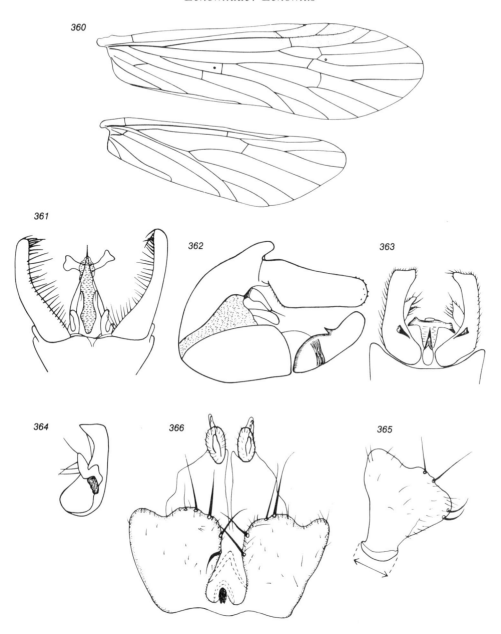

Figs. 360–366: *Ecnomus gedrosicus* Schmid, 1959
360. wings; 361–364. male genitalia
(361. dorsal view; 362. lateral view; 363. ventral view;
364. gonopod seen on median face and somewhat dorso-apically);
365–366. female genitalia
(365. ventral lobe of segment IX, lateral view; 366. ventral view).
(361–364: from Botosaneanu, 1963)

longer than outer branch). Inferior appendages, in lateral view, moderately long and narrow, with very obtuse apex, with a characteristic small excision in the middle of their dorsal border, followed distally by a point, which is also quite distinct in dorsal view; seen dorsally, these appendages are beak-like, turned mediad at their tips, and at the level of the above mentioned excision there arises from the gonopod a narrow annex appendage, obliquely directed upwards and furnished with) long setae. Phallic complex with: (a) a main, sclerotized part, broad basally but suddenly tapering and extremely pointed apically; (b) a membranous part showing apically (when the phallic complex is completely protruded) a pair of lateral appendages bifid at apex; (c) a pair of short titillators united at base and without sharp tips.

Female genitalia (described here for the first time) not forming an ovipositor. Most distinctive for *E. gedrosicus*, compared to *E. galilaeus*, are the lobes of sternite VIII; these are very large, their distal margin very sinuous and furnished with a few long and strong setae, their base considerably shorter than the distal margin (as seen especially in lateral view, where these lobes appear to be pedunculate).

Remarks: According to Schmid (1959:698), this species "... is probably not very near any other species, but it is to be placed not far from *tenellus*". There are differences between the original description from Iran and my own observations on Israeli specimens, pointing to the existence of some variability affecting the size, the venation, and possibly also some male genital characters (but this has yet to be verified).

Distribution: *E. gedrosicus* was described from Shush, in south-west Iran. It was mentioned from the following localities in Israel: Hula, in Upper Galilee (1); various stations around Lake Kinneret, the Jordan River, the Yarmouk River (7); Tannuria, 'Ein Tinna, and Wadi Jalabina (or Gelabina), in the Golan Heights (18). The altitudes range from ca. 575 m a.s.l. (Tannuria) to some 200 m below sea level around Lake Kinneret. It is almost certain that it will be caught at suitable localities in Iraq, Syria, Jordan, and perhaps in Lebanon, too.

Ecological Notes: We do not know anything about the habitat of the species in the Zagros Mountains (Iran). Like so many *Ecnomus* species, *gedrosicus* is certainly an inhabitant of lakes and other bodies of stagnant water; but localities like Wadi Jalabina, a stream in the Golan Heights, show that it can also inhabit running water courses. I consider mentions from the "Jordan River" and "Yarmuk River" (Gasith & Kugler, 1973:59) as not very explicit: does the species really belong to the fauna of these two rivers, or did the specimens caught there originate from Lake Kinneret, for instance? I am unable to answer this question. Adults were caught from March to November; according to light trap data from Lake Kinneret, the species shows two main yearly peaks of population size, one in the spring and the other in the summer or autumn, pointing to the existence of two generations a year.

Family PSYCHOMYIIDAE Curtis, 1835

Br. Ent., Pl. 561 (text) (as Psychomidae)

Type Genus: *Psychomyia* Latreille, 1829

Delicate insects, wings uniformly coloured, venation generally distinct (short setae easily removed). No ocelli. Antennae more slender than in Polycentropodidae, stronger than in Hydropsychidae, at most as long as forewings, with rather long segments, scapus only slightly stronger than the other segments. Maxillary palpi (both sexes) with 5 segments, 1st very short, 2nd longer, 3rd inserted apically (not subapically) on 2nd whose apex lacks stiff setae, last segment long, annulate and flexible. Spurs mostly 2, 4, 4. Tibiae and tarsi of intermediate legs sometimes dilated, sometimes not. Wings elongate, moderately narrow, hindwings narrower than forewings and more pointed (lanceolate); venation slightly reduced in forewings (f 1 absent, discoidal cell short, thyridial cell very small, with basal position and without contact with the longer median cell); in hindwings venation more or less strongly reduced, f 1 absent, discoidal cell open. Vth abdominal sternite without appendages. Male genitalia simple or complex, completely distinct from that of Polycentropodidae; superior appendages well developed, gonopods always bisegmented. Female always with ovipositor, which may be very long.

Key to the Genera of Psychomyiidae in the Levant

1. Third segment of maxillary palpus longer than 2nd. Intermediate legs of female not dilated. **Tinodes** Curtis
 – Third segment of maxillary palpus shorter than (or, at most, as long as) 2nd. Intermediate legs of female dilated 2
2. Forewing rounded at apex. Hindwing without distinct projection of costal border. Male with long gonopods. Female with long ovipositor. **Lype** McLachlan
 – Forewing more pointed. Hindwing with distinct projection of costal border. Male with short gonopods. Female with short ovipositor. **Psychomyia** Latreille

Genus PSYCHOMYIA Latreille, 1829

In: Cuvier, *Règne animal* ed. 2, 5:263 (as *Psychomyie*)

Type Species: *Psychomyia annulicornis* Pictet, 1834, which is a synonym of *P. pusilla* (Fabricius, 1781); subsequent designation by Ross, 1944.

Diagnosis: Small insects of uniform and not very dark colour. Maxillary palpi with 3rd segment shorter than 2nd or 4th. Spurs 2, 4, 4. Tibiae and tarsi of intermediate legs (female) dilated. Wings (especially hindwings) narrow and with apices more pointed than in related genera; in the middle of costal border of hindwings, a very distinct projection. In forewings: f 2, 3, 4, 5 present (f 4 petiolate), very short discoidal cell, longer median cell. In hindwings: f 2, 3, 5 (f 3 sometimes absent; when present,

173

it is petiolate). Male genitalia: sternite IX stout, perfectly distinct; tergite IX curiously fused, partly or entirely, with the superior appendages, which are very elongated, being the most conspicuous part of the genitalia; segment X, and thus intermediate appendages, generally vestigial or absent (there is much incertitude in the descriptions concerning this point); inferior appendages small, with short coxopodites adjacent on median line and set in sternite IX, and with longer and more complex second segments; phallic apparatus roughly resembling a vertical cylinder turned backwards in its distal part and capitate at the end, this variously shaped "head" being possibly the aedeagus. Female genitalia: Segment VIII normal, segments IX + X forming a very short ovipositor (IX broadly cleft ventrally; X with one pair of cerci).

General Remarks: *Psychomyia* is a rather small genus, with some 20 known species; most of them inhabit Eurasia, belonging either to the Palaearctic or to the Oriental Region, a few live on the border between them, and only 3 species are Nearctic. The species are quite diverse, especially with regard to certain characters of venation, but also to some male genital characters; nobody has attempted to revise the genus. Most species are inhabitants of larger water courses, neither very rapid nor very cold; but stagnant waters are also inhabited, and some species (like *P. pusilla*, the unique Levantine representative) are good examples of ubiquitous elements. Larvae construct sinuous tunnels fixed to the substrate, while the pupae are protected by elliptical pupal cases, both typical for the Psychomyiidae.

Psychomyia pusilla (Fabricius, 1781)

Figs. 367–372

Phryganea pusilla Fabricius, 1781, *Spec. Ins.*, 1:392.
Psychomia (sic!) *pusilla* —. Hagen, 1858, *Stettin. ent. Z.*, 19:121.
Psychomyia pusilla —. McLachlan, 1878, *A Monographic Revision & Synopsis*, pp. 426–427.

Wing expanse varying, in 11 ♂ specimens from 'Ein el Sultan, from 6.7 to 9 mm; in 72 males from Naḥal Meshushim, from 7.8 to 9.6 mm; ♀ always larger: ca 10.8–11.8 mm. A pale, uniformly coloured insect: generally pale brown, legs yellow, abdomen with an ochreous tint (colour variations from one population to another). Characters of body and wings as in description of genus *Psychomyia* (wing venation: Fig. 367).

Male genitalia: There is no independent tergite IX, this being coalescent with the comparatively enormously developed superior appendages; this complex is much more distally placed than sternite IX, and its sides are strongly protruding basad (lateral view); sternite IX stout, much higher distally than proximally, with sinuous margins and with very deep and broad excision dorsally. Superior appendages strongly elongated, parallel, with blunt apices; their most distinctive character is a large depression on their medio-dorsal side, completely filled with fine spines arising from papillae; distad to this depression, a distinct, black, roughly triangular zone, resembling a strong beak (without being one). It is not clear what part of the genitalia

174

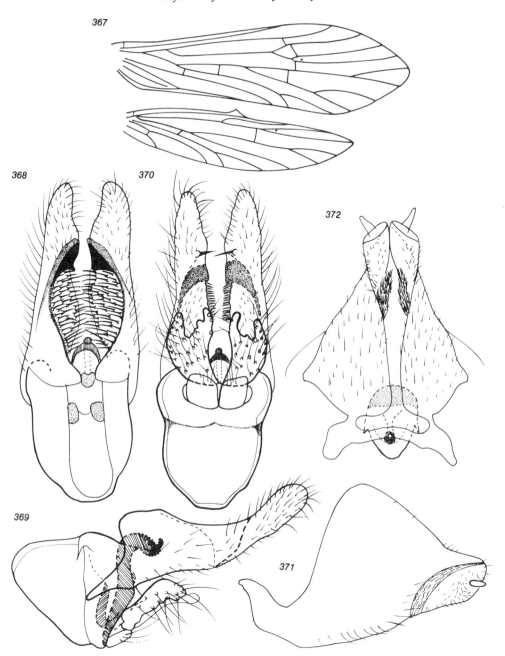

Figs. 367–372: *Psychomyia pusilla* (Fabricius, 1781)
367. wings;
368–370. male genitalia
(368. dorsal view; 369. lateral view; 370. ventral view);
371. female genitalia, essentially segments IX and X, lateral view; 372. the same, ventral view.
(367: from Mosely, 1939a)

belongs to segment X. Inferior appendages comparatively small, bisegmented, basal segment very small, roughly rectangular, distal segment deeply excised apically, this resulting in two similar, moderately broad, not very long, bluntly ending lobes (median lobes slightly turned laterad at apex). Phallic apparatus vertical, but with its distal part curved posteriad in a horizontal plane; its most distinctive part is its apex, resembling a bell in dorsal or ventral view, and a snake's head in lateral view.

Female genitalia: Segments IX and X forming a very short, roughly triangular, laterally compressed ovipositor. Sternite IX completely cleft in its middle, with a pair of short but broad apodemes basally; in the distal part of the median cleft, a pair of stripes densely covered with spines. Segment X short, dorsally and ventrally split on the median line, apical margins oblique, with one cercus on each side. Spermathecal sclerite as in Fig. 372.

Remarks: *P. pusilla* is a typical *Psychomyia*, closely related to two other western Palaearctic species, *P. ctenophora* McL. (Iberian Peninsula) and *P. usitata* McL. (Turkestan, etc). There is an interesting geographic variability, reflected for example, in the male genitalia (see: Schmid, 1959: 767, Pl. 6, figs. 6–8); unfortunately, this has not been thoroughly investigated. A careful comparative study of Levantine specimens from many different localities would possibly also reveal such a variability.

Distribution: *P. pusilla* is a widely distributed species, present almost throughout Europe, and also in North Africa, Asia Minor and northern Iran. In the Levant it is known from Lebanon and from Israel (it will undoubtedly also be found in Jordan, Syria and Iraq). There are many Lebanese localities, either in the catchment of Nahr ed Damour (the river itself, from 260 to 40 m a.s.l., and its tributary Nahr el Hammam), or in that of Nahr el Aouali (the river itself between 380 and ca. 40 m a.s.l., and its tributary Nabaa Joun). The species was also caught on the Litani River at Sir el Gharbii, some 15 km from its mouth, on its tributary Ghozayel at 900–1,000 m a.s.l., as well as at Labwé, a hydrographic complex in the catchment of the Orontes River. There are also rather many localities in Israel and the surrounding areas. In area 1, between ca. 70 and 380 m a.s.l., these are: the hydrographic complex at Tel Dan, the main spring of the River Banias, the rivers Nahal Hermon (Banias), Nahal Senir (Hazbani), and Dan (these three being the main water courses forming the Jordan River) and Hula; in area 18: Wadi Jalabina (or Gelabina) at ca. 100 m a.s.l. and the hydrographic complex at Nahal Meshushim, not far from the northeastern shores of Lake Kinneret and almost in area 7; in this last area is located Deganya A; finally, the following known localities are in area 13: Wadi Auja, at 20 m below sea level, north of Jericho (adjacent to area 12), 'Ein el Sultan in Jericho, Wadi Qilt (10 m a.s.l.), and 'Ein Nueima (130 m below sea level). Special mention should be made for these last two localities, which are possibly the southernmost outposts of the species in the Levant (it was never caught in the Dead Sea Depression south of the northern end of this sea, nor in the Negev).

Ecological Notes: Exclusively found in running water, *P. pusilla* has a very broad ecological spectrum, inhabiting — in the Levant — water courses from the Crenal to the Epipotamal, from at least 1,000 m a.s.l. in the Lebanese mountains, to ca. 150 m below sea level, north of the Dead Sea Depression. It is one of the most ubiquitous and tolerant species in the Province (though, possibly, never inhabiting truly eurythermic, sluggish, and muddy water courses). Despite its mention from Deganya A, I am not convinced that it belongs to the fauna of the Jordan River. In some localities, the populations are enormous, for instance, in the streams at Nahal Meshushim or the spring 'Ein el Sultan. The species is well attracted by lights. In many samples, males clearly outnumber females. On the wing throughout the year, at least in Israel (this possibly does not apply to the higher altitudes in the Lebanese mountains: the Lebanese specimens were caught between April and October).

Genus LYPE McLachlan, 1878
A Monographic Revision & Synopsis, pp. 409, 422

Type Species: *Anticyra phaeopa* Stephens, 1836; subsequent designation by Ross, 1944.

Diagnosis: Small or medium-sized, generally dark insects. Maxillary palpi with 3rd segment shorter than (or as long as) 2nd. Spurs: 2, 4, 4. Intermediate legs of female with distinctly dilated tibiae and tarsi. Both wings moderately broad, without pointed apex, hindwings with costal margin sinuous and without projection. In forewing f2–5, in hindwing f 2, 3, and 5, and discoidal cell either open or closed with an indistinct cross-vein. Male genitalia: sternite IX well developed, tergite IX mostly coalescent with segment X, but partly free as a narrow "dorsal plate"; segment X stout, simple (no intermediate appendages), forming a roof above the phallus; superior appendages simple, elongated, oval; inferior appendages bisegmented, coxopodites short, harpagones much longer, narrow, simple, forming a forceps; phallic apparatus proximally pedunculate (tubular), distally dilated, with a pair of endothecal spines and an apex (aedeagus) generally concave anteriorly. Female genitalia (*L. reducta* Hag.): Segments VIII, IX and X constitute a long ovipositor; segment VIII not divided into tergite and sternite, proximal margins laterally with deep, rounded excisions; segment IX very long, medio-ventrally with narrow cleft, with a pair of short apodemes; segment X small, bilobed and with pair of slender cerci having a ring-like structure.

General Remarks: *Lype* is a very small genus, comprising no more than 8 known species: 4 European or western Palaearctic (two of them with very restricted distribution), 1 in North America, 1 in Ceylon, 1 in India and Burma and 1 African (Ruwenzori). All the known species are morphologically very similar. The presence of a second European species in the Levant is not excluded: *L. phaeopa* (Stephens), is also mentioned from northern Iran. There are several minute, partly unreliable, characters (venation or male genitalia) distinguishing *L. phaeopa* from *L. reducta*, but there is also a good distinctive character for the males: in *L. phaeopa* the apex (aedeagus) of the phallic apparatus is clearly bifurcate in dorsal view, not rounded (Fig. 380), although it is rounded in an Italian subspecies. A second distinctive character of *L. phaeopa* is the "dorsal plate", less slender than in *L. reducta* and more distinctly angular (i.e., apically curved downwards) in lateral view. The larvae and pupae of *Lype* are ethologically similar to those of the other Psychomyiidae; the species are creno- and rhithrobionts.

Lype reducta (Hagen, 1868)
Figs. 373–379

Psychomyia reducta Hagen, 1868, *Stettin ent. Z.*, 29:264–265.
Lype reducta —. McLachlan, 1878, *A Monographic Revision & Synopsis*, p. 424.

Wing expanse, in specimens from Tel Dan, Israel: ♂ 8.7–9.6 mm, ♀ ca 11.4 mm (probably also larger). Body dark brown; antennae rather distinctly annulate; legs very pale; membrane of wings brownish, forewings with very thick covering of easily removed, mainly golden setae (but the general appearance is dark, especially in fresh, dry specimens, and the insects have sometimes been described as black, or blackish). In anterior wing, f3 generally longer than its footstalk, f4 either sessile or only with minute footstalk; in posterior wing f3 much longer than its footstalk, discoidal cell closed with a very indistinct cross-vein.

Male genitalia: From beyond tergite VIII, a long and slender "dorsal plate" formed by segment IX protrudes; this is sometimes almost straight (as in Fig. 375), sometimes gently curved downwards; in lateral view, it is separated by a deep excision from segment X, which is, in dorsal view, very slightly notched apically in its middle. Superior appendages very long, oval, quite simple, moderately blunt at apex. Inferior appendages slightly longer than superior appendages, bisegmented; coxopodite short, stout, distal margin oblique; harpago considerably longer, very simple, in lateral view narrow, slightly tapering towards the blunt apex and very slightly curved upwards; in dorsal or ventral view, the harpagones are regularly curved, distinctly capitate, and each with a beak directed mediad. Phallic apparatus pedunculate basally, then strongly dilated, this dilated part trough-shaped, anteapically with a pair of black spines; its apical part (aedeagus?) is directed upwards, concave anteriorly (spoon-shaped), and terminally slightly rounded, not bifurcate.

Female genitalia: see Figs. 378–379 and description of genus *Lype*.

Distribution and Ecological Notes: A typical European species, *L. reducta* is known from many parts of the Continent, but not from its northern regions; also known from Asia Minor and from Tunisia and Morocco. In the Levant the oriental and meridional limits of its distribution are probably reached. It was caught in several localities in Lebanon: Ghozayel, a tributary of the Litani, at Anjar Chamsine (900–1,000 m a.s.l.); Nahr el Aouali at Jdaidet ech Chouf (in the upper course of this coastal river, ca. 700 m a.s.l.); and on several springs and streams in the Aouali basin (800–850 m a.s.l.). In Israel, the species is possibly restricted to Upper Galilee (1) where only two localities are known: the hydrographic complex at Tel Dan (Tel el Kadi), and Naḥal Layish,

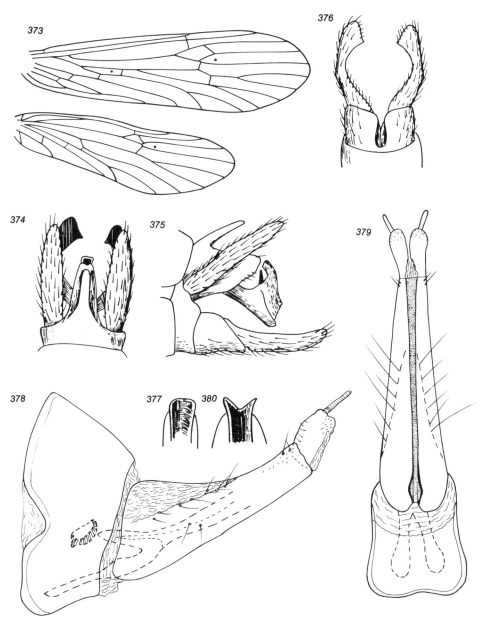

Figs. 373–379: *Lype reducta* (Hagen, 1868)
373. wings; 374–377. male genitalia
(374. dorsal view; 375. lateral view; 376. ventral view;
377. apex of phallic apparatus — or "aedeagus", dorsal view);
378–379. female genitalia (378. lateral view; 379. ventral view)

Fig. 380: Apex of phallic apparatus — or "aedeagus", dorsal view,
in *Lype phaeopa* (Stephens, 1836). (From Mosely, 1939a, except 378–379)

179

a streamlet at Bet Ussishkin, in the catchment of the River Dan; these two localities are at 175–200 m a.s.l. This species is a crenobiont and rhithrobiont also inhabiting large streams belonging to the Hyporhythral. It is seldom caught in large numbers, except at Tel Dan. On the wing in the Levant practically all the year round.

Genus TINODES Curtis, 1834
Phil. Mag., 4:216

Type Species: *Tinodes lurida* Curtis, 1834 (by monotypy); this is a synonym of *Phryganea waeneri* Linnaeus, 1758.

Diagnosis: Small to medium-sized insects (Fig. 381), forewings very setose and uniformly coloured (from light brown or beige to black in various species). Antennae shorter than forewings. Maxillary palpus with 3rd segment longer than 2nd. Spurs: 2, 4, 4. Intermediate legs of female not dilated. Wings: Fig. 382. Forewings moderately narrow, with elliptical apex, setae very dense but easily removed leaving venation distinct; at root of SR a round "glabrous cell", not always very distinct; discoidal cell short, irregularly angular, median cell much longer and narrower, thyridial cell smaller than the median, and much more basal; f 2, 3, 4, 5 present (2: long and sessile; 3 and 4: stalked). Hindwings narrower and shorter, costal margin slightly concave in its distal part, but without a sharp projection; discoidal cell open, f 2, 3, 5 present (3: sessile). Male genitalia complex and characteristic (interpretation of their parts sometimes difficult, my interpretation sometimes divergent from that of other authors). Ventral part of segment IX well developed, sclerotized, variously shaped; its upper proximal angles are continued, on each side, by a sclerite ("sclerite of the genital chamber") directed upwards to join the root of the phallic complex. The most dorsal part of the genitalia is a triangular piece, mainly membranous (segment X vestigial), with a narrow sclerotized frame (tergite IX?); it does not offer good specific characters. Nor are the long, simple, often slender superior appendages useful in distinguishing species. It may be mentioned that the root of tergite IX (?) and that of the superior appendage converge towards the limit between the "sclerite of the genital chamber" and sternite IX. Inferior appendages always large and complex, offering excellent specific characters; coxopodite much longer than harpago, very variable in shape, often with various conspicuous projections; harpago simple, much smaller, rooted subapically on the median side of the coxopodite; finally, a large, complex, variously shaped internal (median) armature between the two gonopods, obviously forming a "guide" for the phallic complex situated above it, and with a long, proximal, horizontal apodeme. What I have named "phallic complex" is a horizontal structure having a remarkably high position in the genitalia; proximally there is often a sclerotized sheath, often furnished with very conspicuous spines and/or spurs; the phallus proper is mainly membranous, and the aedeagus is seen inside it. Female genitalia forming a very long, morphologically complex ovipositor in which segments VIII, IX, and X are involved, segment IX being by far the longest. Tergite VIII

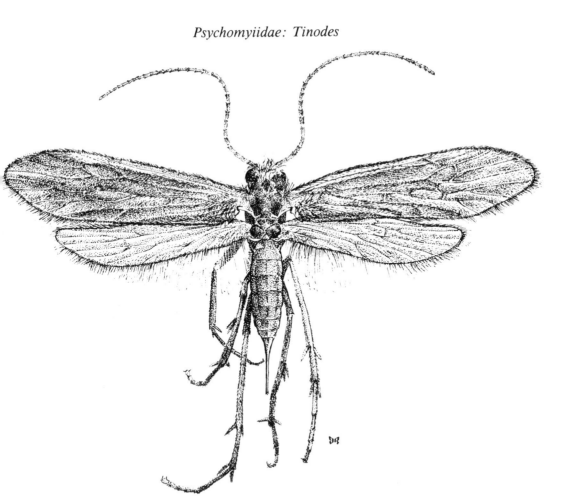

Fig. 381: *Tinodes negevianus* Botosaneanu & Gasith, 1971, female

normally developed; pleura VIII broad or narrow, sometimes with a deep pit (*d* in Fig. 401) in which the distal points of the male coxopodite are anchored during copula; sternite VIII less developed, either not interrupted medially, or divided into two "ventral lobes" (*c* in Fig. 402) united only by a narrow strip (behind these lobes, the harpago is anchored during copula). Segment IX extremely elongated, gradually tapering, with distinct medio-ventral cleft (*e* in Fig. 402) in which tergite IX + segment X of the male are inserted during copula; this cleft is continued basally (ventral view) by a wide opening with rounded margins (*a* in Fig. 402); the margins of this opening are sclerotized and actually form (lateral view: *a* in Fig. 401) a pair of sclerites apparently continuing the lower angle of the root of segment IX, directed cephalad and then upturned; the upper angle of the root of segment IX is sometimes only slightly produced, but at least in one species (*b* in Figs. 401–402) developed into a long, strong sclerite which is trough shaped (concave ventrally). Segment X elongated but very short compared with segment IX. Spermatheca in several species weakly chitinized, but in one of them with strong distal sclerite.

181

General Remarks: A large genus, with probably some 140 known species, most of them with restricted distributions; almost cosmopolitan, but not present in Central or South America; one species in Australian; without any doubt, the largest number of species — some 75 known, several more to be discovered — inhabit the western Palaearctic (*sensu lato*, including also the adjacent parts of the Oriental Region), especially around the Mediterranean. It is a very interesting genus, but there is no revision of it; species fall into species-groups, sometimes distinct, sometimes less so. For the Levant it is one of the most distinctive genera, the 5 known species, belonging to two distinct species-groups, being probably endemic or, at least, quasi-endemic. There is little hope of discovering more species in Israel, but additional ones will possibly be found in northern Lebanon and the limitroph zones of Syria. The species are crenobionts and/or rhithrobionts, often with a tendency to a hygropetric mode of life; larvae and pupae build the sinuous tunnels fixed to the substrate, and, respectively, the cocoons protected by elliptical pupal cases, typical for the family.

Key to the Species of Tinodes in the Levant

♂♂

1. Coxopodite without sharp distal projections   2
 –   Coxopodite with sharp distal projections   3
2. Coxopodite very simple, lower distal angle not projecting; 2 pairs of long spurs on sclerotized sheath of "phallic complex".   **T. kadiellus** Botosaneanu & Gasith
 –   Coxopodite simple, but with lower distal angle produced in a blunt lobe; 4 pairs of long spurs on sclerotized sheath of "phallic complex".   **T. tohmei** Botosaneanu & Dia
3. Coxopodite strongly tapering distad, only its upper distal angle forming a long, sharp "beak"; phallus accompanied by a pair of long, slender appendages ending in a long spur.   **T. negevianus** Botosaneanu & Gasith
 –   Coxopodite not strongly tapering distad, with 3 distal, sharp projections; "phallic complex" proximally with lateral sclerotized walls equipped with variable sets of spines   4
4. Upper 2 of the 3 distal projections of coxopodite equally long, forming a kind of forceps.   **T. caputaquae** Botosaneanu & Gasith
 –   Upper distal projection of coxopodite much stronger and longer than that following ventrally.   **T. israelicus** Botosaneanu & Gasith

♀♀

(This key should be used with caution; the association of *T. tohmei* and *T. kadiellus* is not fully ascertained, although probably correct; the female of *T. israelicus* was not associated: it will probably key near *caputaquae*)

1. Sternite VIII not completely divided into two lobes (although weakly chitinized and emarginated distally).   **T. tohmei** Botosaneanu & Dia
 –   Sternite VIII divided into two "ventral lobes"   2

2.　From dorsal angle of "root" of segment IX, a long, strong, trough shaped sclerite extends cephalad; a deep pit on pleura VIII.　　　**T. caputaquae** Botosaneanu & Gasith

–　Dorsal angle of "root" of segment IX rounded, not followed cephalad by such a sclerite; no deep pit on pleura VIII　　　　　　　　　　　　　　　　　　　3

3.　Small "ventral lobes" of sternite VIII united by a relatively broad chitin strip; "opening" at proximal end of ventral cleft of segment IX, short, almost round.

T. negevianus Botosaneanu & Gasith

–　Small "ventral lobes" united only by an extremely narrow chitin strip; "opening" at proximal end of ventral cleft elongated, oval.　　　**T. kadiellus** Botosaneanu & Gasith

Tinodes caputaquae Botosaneanu & Gasith, 1971

Figs. 382–385, 401–402

Tinodes caputaquae Botosaneanu & Gasith, 1971, *Israel J. Zool.*, 20:109–112. Type Locality: 'En Po'em (Wadi Lamoon), Upper Galilee, Israel. Type deposited in the Zoological Museum, Tel Aviv University.

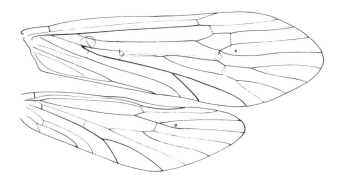

Fig. 382: *Tinodes caputaquae* Botosaneanu & Gasith, 1971, wings

Wing expanse: ♂ ca. 11.5–13 mm, ♀ ca. 15 mm (a large species).

Male genitalia: Sternite IX in ventral view quadrangular; in lateral view roughly quadrangular, too, its lower proximal angle only very slightly protruding, upper proximal angles continued on each side by a rather short and stout sclerite directed upwards and caudad, these two sclerites coalescent terminally where they join the root of the phallus. Inferior appendages with coxopodite, harpago, and complex median (or internal) armature. Coxopodites strongly developed, dorsal and ventral borders almost parallel; upper apical angle protruding in a strong, long, pointed, slightly irregular extension, immediately followed, in a ventral direction, by a quite similar extension (these two appendages somewhat resembling a forceps); a rounded sinus separates this second extension from a short, sharp point formed by the lower distal angle of the coxopodite. Part of the harpago protruding beyond the coxopodite, triangular, obliquely descendent, apex pointed; apices of the two harpagones turned

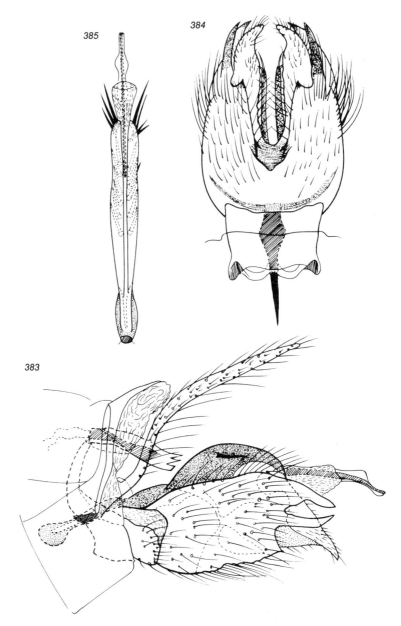

Figs. 383–385: *Tinodes caputaquae* Botosaneanu & Gasith, 1971, male genitalia
(383. lateral view; 384. ventral view; 385. phallic apparatus, dorsal view).
(From Botosaneanu & Gasith, 1971)

mediad. Internal (median) armature having: (a) a long proximal apodeme, straight proximally where it is very broadened (lateral view); (b) a pair of large dorsal processes, resembling two sclerotized wings, dorsally divergent, in lateral view very high basally and then regularly tapering to a slender apical part curved basad; (c) a pair of ventral baculiform processes (seen in Fig. 384), apically with two small spines. Phallic complex very narrow basally, then gradually broadened, with very long lateral sclerotized walls; apically-subapically on the ventral face of each of these sclerites, normally some 3–6 spines of varying length (for variability, see "Remarks"); also on the ventral face of each sclerite, but distinctly more proximally and closer to the median line, a longitudinal row of some 6–7 spines (but see "Remarks" for variability).

Female genitalia (female certainly correctly associated): On pleura of segment VIII a large and deep pit; sternite VIII completely divided into two lobes proximally united by a narrow chitin strip (these lobes with some kind of ventro-median pouch). Superior angle of root of segment IX followed cephalad by a long, strong sclerite, distinct laterally and ventrally, and trough shaped (concave beneath); "opening" at basal end of the medio-ventral cleft on segment IX, elongated-oval.

Remarks: This species belongs to a fairly large group distributed in the Balkan Peninsula, the Aegean Islands, and the Levant, with one species in Spain (group of *raina*, after the earliest named species). It is the sister-species of *T. israelicus* Botosaneanu & Gasith, 1971. As in other species of the genus, some variability in the number of spines on the sclerotized walls of the phallus was noted in different populations. In some Lebanese specimens I have found only 1 spine apically on each side. Malicky (1983:99) provides drawings of *T. caputaquae* (from which part of the Levantine Province?), the phallus being represented with 3 pairs of apical spines, and without proximal spines.

Distribution and Ecological Notes: *T. caputaquae* is known only from the Levant. It is largely distributed in the catchments of two Lebanese coastal rivers, Nahr ed Damour and Nahr el Aouali, and especially in springs and spring-brooks of the Barouk-Niha massif, from ca. 40 m up to ca. 1,000 m a.s.l., but also in larger water courses (N. el Aouali, N. ed Damour, Nahr el Hamman). It is known, moreover, from the upper course of the Orontes (springs and streamlets at Labwé, 1,000 m a.s.l.), and from the tributary Yahfoufa of the Litani River, at Janta (1,100 m a.s.l.). One female specimen was caught at Senir, in the Golan Heights (18). In Israel, the species is known from two localities in Upper Galilee (1): En Poem (or Wadi Lamoon), which is a small, rapidly flowing stream at 550 m a.s.l., and Ein Habis, a spring of Nahal Keziv, at ca. 320 m a.s.l.; a few females present in the collections were possibly caught in some other localities in Upper Galilee. It is interesting that this species was never found at Tel Dan, an excellent potential locality, where the closely related *T. israelicus* was caught. *T. caputaquae* apparently does not extend further south than Galilee. A crenobiont-rhithrobiont with a rather broad altitudinal range, and probably forming large populations. Adults were caught from March to December.

Tinodes israelicus Botosaneanu & Gasith, 1971

Figs. 386–388

Tinodes israelica Botosaneanu & Gasith, 1971, *Israel J. Zool.*, 20: 106–109. Type Locality: Tel Dan (Tel el Kadi), Israel. Type deposited in the Zoological Museum, Tel Aviv University.

Wing expanse, in the two known ♂ Israeli specimens: 7.1–7.7 mm (a small species).

Male genitalia: Sternite IX in ventral view quadrangular; in lateral view roughly quadrangular, too, its lower proximal angle only very slightly protruding, upper proximal angles continued on each side by a long and slender sclerite directed first cephalad, then upwards and caudad, these two sclerites coalescent terminally, where they join the root of the phallus. Inferior appendages: Coxopodite as in *T. caputaquae*, but with an important difference, the strong and long extension of its upper apical angle being followed, in a ventral direction, by a very sharp and distinctly shorter extension (separated by a rounded sinus from the short, sharp point of the lower distal angle of the coxopodite); part of the harpago protruding beyond the coxopodite, broadly triangular, obliquely descendent, with obtuse apex; internal (median) armature roughly similar to that in *T. caputaquae*, but the apodeme is less broadened at its cephalic end. Phallic complex very narrow basally, then gradually broadened, with very long lateral sclerotized walls, each furnished — in the two known specimens from Tel Dan — with only one long and slender terminal spine with terminal part finely sculptured (for variability, see "Remarks").

Female presently not associated; it must be very similar to that of *T. caputaquae*.

Remarks: Like its sister species *T. caputaquae*, *T. israelicus* belongs to the moderately large group of *raina*, distributed in the Balkan Peninsula, the Aegean Islands, and the Levant, with one species in Spain. It will be distinguished from *T. caputaquae* by its smaller size, and by some male genital features, one of them conspicuous and easily seen: the shape of the distal extensions of the coxopodite. On the basis of current information, it is possible to say that the two sister-species apparently never coexist in the same locality. Malicky (1981: 339–340) mentions for several specimens caught at Bcharré, Lebanon, differences exclusively in the armature of spines of the phallus; there are considerably more spines, on each side some 4 distal and some 6 subdistal; impressive as they may appear, such differences merely serve to distinguish different populations (but they are certainly real differences, not due to the severance of some spines in the Israeli specimens).

Distribution and Ecological Notes: Presently known only from two localities: (a) Bcharré (Beshari), in northern Lebanon, at 1500 m a.s.l.; (b) the complex of springs and streams at Tel Dan (or Tel el Kadi), at 200 m a.s.l. in Upper Galilee, Israel (1). Probably a rare species, sporadically distributed, and a crenobiont (crenobiont-rhithrobiont?). Adults caught in May and June.

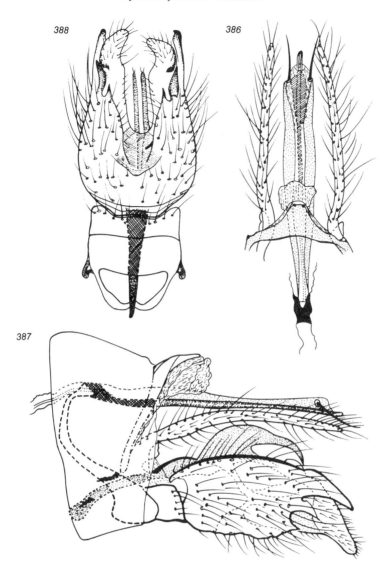

Figs. 386–388: *Tinodes israelicus* Botosaneanu & Gasith, 1971, male genitalia
(386. dorsal view; 387. lateral view; 388. ventral view).
(From Botosaneanu & Gasith, 1971)

Tinodes negevianus Botosaneanu & Gasith, 1971

Figs. 381, 389–391, 403–404

Tinodes negeviana Botosaneanu & Gasith, 1971, *Israel J. Zool.*, 20:115–118. Type Locality:
Naḥal Dawid ('En Gedi, Dead Sea Depression, Israel). Type deposited in the Zoological
Museum, Tel Aviv University.

Wing expanse: ♂ ca 8.2–11 mm, ♀ ca 12.4 mm. Habitus: Fig. 381.

Male genitalia: Sternite IX, in ventral view, characteristically trapezoid with lateral margins sinuous, posterior border thickened and with deep angular excision, anterior border with small median extension bordered by sinuses; laterally it is roughly triangular, proximal lower angle slightly and bluntly produced; upper proximal angle continued on each side by a short, stout, straight (oblique) sclerite (these sclerites reaching the root of the phallus). Coxopodite of inferior appendage very high at base, then strongly tapering, its apical part being a long and narrow, sharp "beak" slightly curved downwards and mediad, with ventral margin separated from the very oblique ventral border of the coxopodite by a sinus. Part of the harpago protruding beyond the coxopodite, with a dorsal lobe (horizontal, broadly digitiform, with blunt apex reaching the tip of the "beak" of the coxopodite) separated by a small sinus from a much smaller ventral lobe – a triangular point. Internal (median) armature of the inferior appendages small, but with long apodeme sagitally flattened, with proximal end broadened and slightly turned to the left; the most characteristic part of this armature is a pair of very long and slender appendages, slightly broadened in their anteapical parts, each having a long, strong spur inserted apically and finely sculptured (there is also a minute anteapical seta); these appendages, acting probably as titillators, accompany the phallus but do not belong to the phallic apparatus (in Fig. 389 they were represented together with the phallus). Phallic complex devoid of armature of spines.

Female genitalia (female certainly correctly associated): Sternite VIII divided into two small lobes with rounded angles, united by a relatively broad chitin strip. No long sclerite following cephalad the blunt superior angle of root of segment IX; medio-ventral cleft of segment IX markedly broadened distally, where the segment is completely open (perhaps not a truly reliable character); at base of this cleft, an "opening" which is rather short, almost round.

Remarks: *T. negevianus* belongs to the same species-group (*raina*) as *T. caputaquae* and *T. israelicus*, but it certainly has a much more isolated position within this group.

Distribution and Ecological Notes: This species, too, is a potential endemic element of the Levant. It was found in only one Lebanese locality: the spring Nabaa Aazibi and its streamlet, in the basin of Nahr Aaray (tributary of Nahr el Aouali), at ca. 1,000 m a.s.l. To the south of Lebanon, the localities are rather numerous, *T. negevianus* being almost certainly the sole *Tinodes* extending southwards to the semi-desert or desert zones; in Samaria (6) it was caught at Shechem (Nablus), at ca. 850 m a.s.l. as well as on Wadi Faria at 280 m below sea level; in the Judean Hills (11) it was found in large numbers in two springs at Wallaja, near 'Amminadav, at 650 m a.s.l.; finally, it is known from 'En (Naḥal) Mishmar and from Naḥal Dawid ('En Gedi) in the Dead Sea Area (13), at ca. 350 below sea level. The apparent gap between Lebanese and Israeli localities cannot presently be explained. *T. negevianus* is (almost) a crenobiont, with a broad altitudinal range (from at least 1,000 m a.s.l. to 350 m below sea level). At most of the localities adults were caught in March and April, but some were caught in June–August or in February (at Naḥal Dawid).

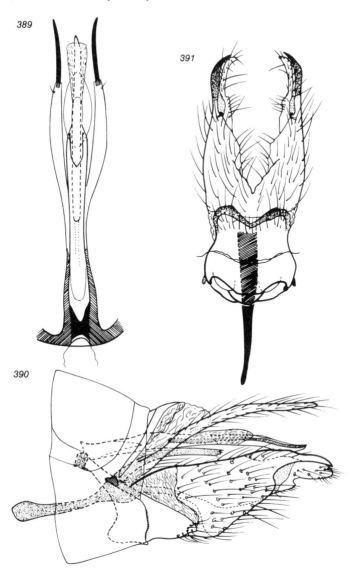

Figs. 389–391: *Tinodes negevianus* Botosaneanu & Gasith, 1971, male genitalia
(389. phallic apparatus, on its sides with pair of appendages belonging to the
internal armature of inferior appendages; 390. lateral view; 391. ventral view).
(From Botosaneanu & Gasith, 1971)

Tinodes kadiellus Botosaneanu & Gasith, 1971
Figs. 392–396, 405–407

Tinodes kadiella Botosaneanu & Gasith, 1971, *Israel J. Zool.*, 20:112–115. Type Locality: Tel Dan (Tel el Kadi), Israel. Type deposited in the Zoological Museum, Tel Aviv University.

Wing expanse: ♂ type from Tel Dan, Israel: 7.7 mm; 1 ♂ from Nabaa Ras el Ain, Lebanon: ca. 10.3 mm; 1 ♀ from Tel Dan: ca. 8 mm.

Male genitalia: In ventral view, sternite IX resembles a roughly triangular shield, proximally pointed or blunt, distally more or less deeply excised (compare Figs. 393 and 396); laterally stout, distal end truncate, proximal lower angle sharp, proximal upper angles continued, on each side, by a short and broad, not curved, oblique sclerite, these two sclerites independently reaching the root of the phallus. Coxopodite of the inferior appendage very stout, twice as high at its proximal end as at its distal end, upper limit convex; distal end vertical, devoid of any projection; in ventral view the two coxopodites form a roughly circular complex. Harpago very simple: a broadly digitiform segment with blunt apex, directly inserted at apex of the coxopodite; a very characteristic strong brush of rigid setae is laterally inserted on the harpago near its base. Internal (median) armature of inferior appendage with: a) a long proximal apodeme sagitally flattened, with proximal end strongly turned to the left; b) only one pair of processes (ventral processes lacking), resembling very large sclerotized wings, divergent dorsally, in lateral view with dorsal margin convex, ventral margin deeply excised, distal margin divided by an excision into a shorter dorsal and a longer ventral lobe. Phallic complex: proximal sclerotized part relatively short, stout, cleft only distally on dorsal as well as on ventral side; on dorsal side, and subapically, are inserted two pairs of very long and strong, similar spurs, sculptured along most of their length; from this proximal sclerotized part a long, slender, only slightly chitinized (partly membranous) tube protrudes.

Female genitalia (the female is not associated with absolute certainty): Sternite VIII divided into two small, triangular lobes united only by an extremely narrow chitinous line. No long sclerite following cephalad the blunt superior angle of root of segment IX; ventro-median cleft of segment IX at its base with an elongated-oval "opening". The spermatheca — its posterior part probably — is particularly well sclerotized compared with other species, the sclerite forming a strongly curved "half-ring" (concave above), strongly excised posteriorly, with rounded lateral lobes.

Remarks. *T. kadiellus* belongs to a compact western Palaearctic group of species, that of *unicolor*. There is a slight variability, some male Lebanese specimens differing from those from Tel Dan by their larger size, a slightly differently shaped IXth sternite, and perhaps a slightly more slender harpago with dorsal margin, in lateral view, slightly (more) emarginate.

Distribution and Ecological Notes: In Lebanon, known from two localities in the catchment of Nahr el Aouali, Barouk-Niha massif: Ras el Ain, at 850 m a.s.l., and Nabaa Aazibi, at 990 m a.s.l. The only known Israeli locality is the complex of springs

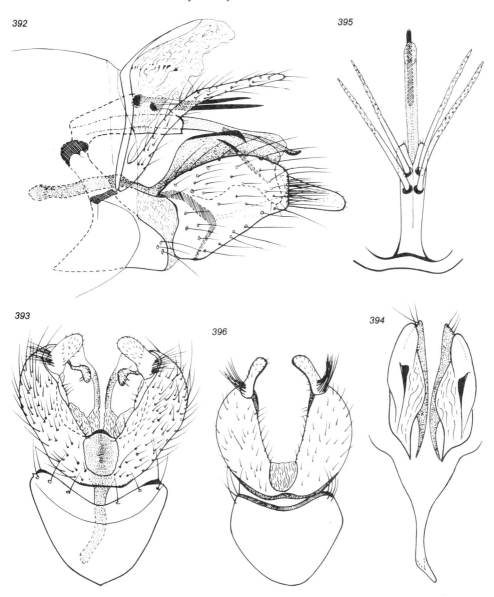

Figs. 392–396: *Tinodes kadiellus* Botosaneanu & Gasith, 1971, male genitalia
392–395. holotype from Tel Dan, Israel
(392. lateral view; 393. ventral view;
394. internal armature of inferior appendages, dorsal view;
395. phallic apparatus, dorsal view);
396. specimen from Nahal Ras el Ain, Nahr el Aouali, Lebanon
(sternite IX and gonopods, ventral view).
(From Botosaneanu & Gasith, 1971, except 396)

and streams at Tel Dan in Upper Galilee (1), perhaps the southernmost point in the species' distribution. *T. kadiellus* is, moreover, mentioned from Rhodes and Cyprus (I did not see specimens from these islands); it was reported from Asia Minor, too, but is absent from the list by Malicky & Sipahiler (1984). This is a true crenobiont, all the known localities being springs or spring-brooks, and an apparently very rare species. Adults are on the wing from April to June.

Tinodes tohmei Botosaneanu & Dia, 1983
Figs. 397–400, 408–409

Tinodes tohmei Botosaneanu & Dia, in: Dia & Botosaneanu, 1983, *Bull. zool. Mus. Amsterdam*, 9(14): 132–133. Type Locality: Nabaa Salman stream (Ouâdi Ras el Mâ), near Harêt Jandal ech Chouf, Lebanon. Type deposited in the Zoological Museum, University of Amsterdam.

Wing expanse: ♂ 11.2–11.8 mm, ♀ 11.4–12 mm (a relatively large species).
Male genitalia: Sternite IX roughly rectangular in lateral and in ventral view, distal margin vertical in lateral view, lower proximal angle slightly protruding, upper proximal angle produced in slender and nearly vertical sclerites reaching the root of the phallus. Coxopodite of inferior appendage higher basally than distally, inferior border emarginate, lower proximal angle forming a massive triangular projection, and with a characteristic, obtuse and slightly blackened lobe strongly projecting from the lower distal angle. Harpago arising from above the mentioned lobe; it is a simple, elongate segment with obtuse apex; the two harpagones are distinctly turned mediad. Internal (median) armature of the gonopods with long and slender apodeme with apex free and asymmetrically turned to the right; the dorsal processes are enormously developed, upwards divergent, with posterior dorsal part forming on each side two strong, connected thickenings, and with inferior margins forming the longitudinal "swellings" represented in Fig. 398; there are no ventral processes. Phallic complex with basal sclerotized part deeply cleft distally; four pairs of long and slender spurs, distally with sculptured surface, are laterally inserted on this sclerotized sheath, their exact position being shown in Fig. 400.
Female genitalia (female almost certainly correctly associated): Sternite VIII not divided into two "ventral lobes" (although its medio-distal part is only weakly chitinized), and emarginate (proximal limit strongly sclerotized). No long sclerite following cephalad the blunt superior angle of root of segment IX; cleft of segment IX followed at base by an "opening" which is rather short, roughly rhombic.
Remarks: *T. tohmei* belongs to the compact Palaearctic *unicolor* species-group; it is a very distinctive species, possibly more closely related to *T. kadiellus* Botosaneanu & Gasith, *T. erato* Malicky (from Greece), and *T. amadai* Schmid (from Iran).
Distribution and Ecological Notes: Of all the Levantine *Tinodes*, this is the northernmost species, being possibly absent from Israel. All the known localities are in the catchment of Nahr el Aouali, a small coastal river of Lebanon: the stream Ouadi

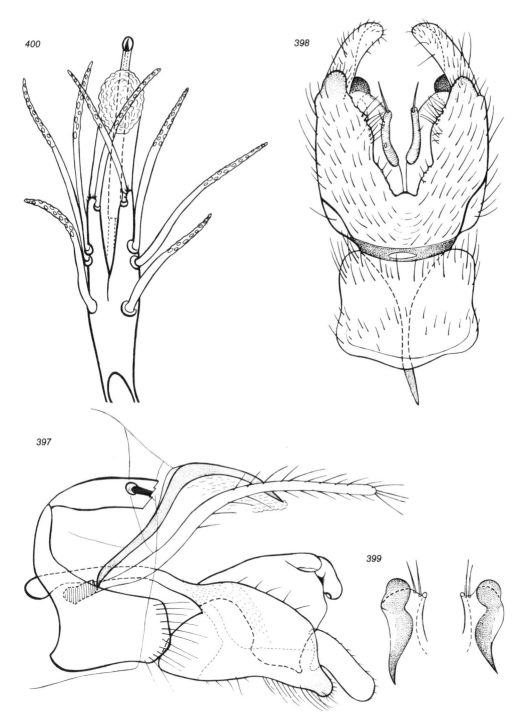

Figs. 397–400: *Tinodes tohmei* Botosaneanu & Dia, 1983, male genitalia
(397. lateral view; 398. ventral view; 399. dorsal processes
of internal armature of inferior appendages, dorsal view;
400. phallic apparatus, dorsal view).
(From Dia & Botosaneanu, 1983)

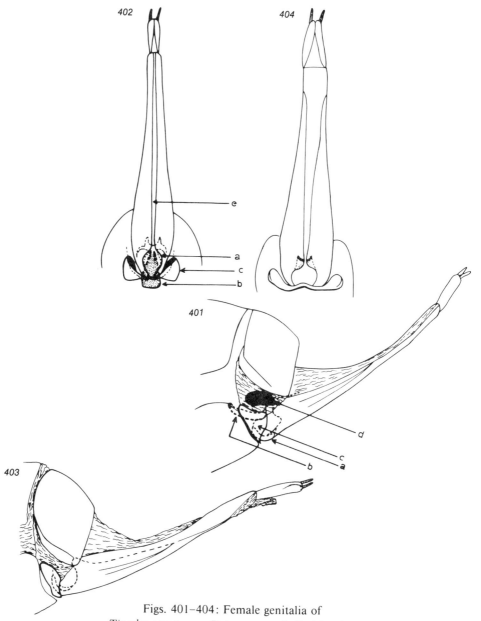

Figs. 401–404: Female genitalia of
Tinodes caputaquae Botosaneanu & Gasith, 1971
(401. lateral view; 402. ventral view)
and of *T. negevianus* Botosaneanu & Gasith, 1971
(403. lateral view; 404. ventral view)
In Figs. 401–402: a –basal widening of medio-ventral cleft on segment IX;
b – sclerite of upper angle of segment IX root;
c – lobes of sternite VIII; d – pit on segment VIII pleura;
e – medio-ventral cleft on segment IX

Ras el Mâ fed by the spring Nabaa Salman, the karst spring Nabaa Mourched, the spring Nabaa Bâter ech Chouf and its emissary stream, and Nahr el Aouali near Jdaidet ech Chouf (700–900 m a.s.l.). A crenobiont-rhithrobiont, caught in rather large numbers in May and June.

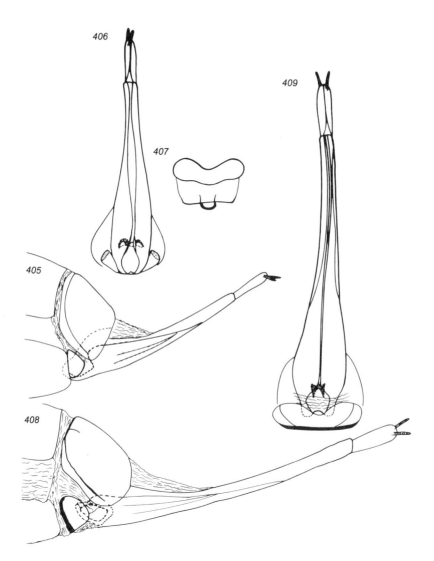

Figs. 405–409: Female genitalia of
Tinodes kadiellus Botosaneanu & Gasith, 1971
(405. lateral view; 406. ventral view;
407. spermathecal sclerite, dorsal view)
and of *Tinodes tohmei* Botosaneanu & Dia, 1983
(408. lateral view; 409. ventral view).
Association of these females not quite certain!

Family BRACHYCENTRIDAE Ulmer, 1903

Abh. Verh. naturw. Ver. Hamburg, 18:85 (as Brachycentrinae)

Type Genus: *Brachycentrus* Curtis, 1834.

Ocelli absent. Antennae slender, at most as long as forewing; scapus as long as or slightly longer than head. Maxillary palpi: 3 segments in male (setose, adpressed to head), 5 segments in female (normal). Spurs short, various formulae, only 2 or 3 spurs on hindlegs. Wings moderately broad, regularly elliptical at end, hindwings generally only slightly narrower than forewings. Neuration slightly simplified, displaying sexual dimorphism (apical forks in forewing: 1, 2, 3, 5 in male, 1–5 or 1, 2, 3, 5 in female; in hindwing: 1, 5 or 1, 2, 5 in male, 1, 2, 3, 5 or 1, 5 in female); forewing with closed and small discoidal cell, closed and long thyridial cell, open median cell; hindwing with discoidal cell open or closed. Green pigment in abdomen. Male genitalia simple, gonopods simple, uni- or bisegmented. Female genitalia with very strongly developed sternite VIII involved in the formation of an apical pouch for the retention of the egg mass. One genus represented in the Levant: *Brachycentrus* Curtis.

Genus BRACHYCENTRUS Curtis, 1834

Phil. Mag., 1834, 4:215

Type Species: *Brachycentrus subnubila* Curtis, 1834 (by monotypy).

The genus *Oligoplectrum* McLachlan, 1868 (type species: *Dasystoma pulchellum* Rambur, 1842 — which is a synonym of *Phryganea maculata* Fourcroy, 1785; by implicit subsequent designation by McLachlan, 1876), to which the unique known Levantine species belongs, is considered by Schmid (1980) as synonymous with *Brachycentrus*; for Flint (1984) it is a subgenus of *Brachycentrus*, and I shall adopt here this point of view.

Diagnosis: This short diagnosis, with the exception of the spur formula, applies to most, or all, members of genus *Brachycentrus*. Medium-sized insects, males of slender appearance, females heavier. Green pigment in the abdomen (the same is true for larvae and pupae). Maxillary palpi with 3 segments in the male and 5 in the female. Spurs 2, 3, 3 (in sg. *Brachycentrus*) or 2, 2, 2 (in sg. *Oligoplectrum*). Moderately broad wings (hindwings with almost parallel costal and postcostal borders); in forewing R1 quite distinctly bent anteapically; important sexual dimorphism in the apical forks: in the male, forewing with f 1, 2, 3, 5 and hindwing with f 1, 5, in the female, forewing with f 1, 2, 3, 4, 5 and hindwing with f 1, 2, 3, 5. Male genitalia of relatively simple structure: superior appendages basally coalescent and forming a plate with markedly dorsal position; below this, the simple, generally bilobed segment X; inferior appendages unisegmented, adpressed to segment IX in their basal part, with free distal part offering some specific characters (but the species are generally difficult to distinguish); phallus tubular, of simple structure. Segments of the female abdomen highly telescopic, with an apical pouch for retention of the egg-mass; the most

196

distinctive character of the genital segments is the enormous sternite VIII (or: VIII + IX ?) playing an important role in the formation of the apical pouch (in some species, sternite VII is also modified).

General Remarks: *Brachycentrus* is a genus of medium importance; it is best represented in the Nearctic, with 13 known species (Flint, 1984); most of the remaining 9 probably valid species are Palaearctic, only a few of them being Oriental. It is possible that sg. *Oligoplectrum* has only one species — the European *maculatus* here described; a species from the Ussuri possibly also belongs to this subgenus. Most of the species — and *B. (O.) maculatus* is no exception — are inhabitants of larger water courses with stony substrates (Hyporhithral, Epipotamal). The larvae of many *Brachycentrus* species build quite typical prismatic (quadrangular in section) cases from vegetable material; *B. (O.) maculatum* is an exception, its larval case being conical, very slender, very narrow at the posterior end, built from sand or sand with vegetable materials, having a variegated appearance.

Brachycentrus (Oligoplectrum) maculatus (Fourcroy, 1785)
Figs. 410–417

Phryganea maculata Fourcroy, 1785, *Entom. Paris*, p. 355.
Oligoplectrum maculatum —. McLachlan, 1876, *A Monographic Revision & Synopsis*, p. 258.

In Lebanese specimens, wing expanse 10–11.8 mm in the ♂, and 10.6–14 mm in 3 ♀♀ (size smaller than in Central European populations). General appearance of male more slender than that of the heavier females. Head and thorax almost uniformly dark brown (blackish); green pigment in the abdomen. Antennae dark brown, serrate, distinctly shorter than forewings, with large scapus. Legs slender; spurs 2, 2, 2 in both sexes, spurs on forelegs very short. Hindwings much shorter than forewings, both wings with very short, dark setae, venation fairly distinct. Forewings moderately broad, with parabolic apex, R1 with very distinct bend in the pterostygmal zone, discoidal cell closed, with sexual dimorphism in the forks: f 1, 2, 3, 5 in the male, and 1, 2, 3, 4, 5 in the female. Hindwings broad (costal and postcostal margins almost parallel), apex obtuse, SC and R1 very close to each other for most of their length, discoidal cell open, marked sexual dimorphism in the forks: f 1, 5 in the male, and 1, 2, 3, 5 in the female.

Male genitalia: Distally from the middle of sternite VII is an elongated, narrow, quadrangular plate with smooth surface. All parts of the genitalia are simple. Segment IX with very characteristic appearance in lateral view: sternite strongly protruding proximad, separated by a deep sinus from the broad median part of the segment, tergite moderately narrow. Superior appendages forming a plate with a clear dorsal position; they are short and broad, coalescent on the median line in their basal half, then broadly separated by a triangular cleft. Below them, the very simple, stout, segment X, apically forming two rounded lobes with minute spines. In lateral view,

197

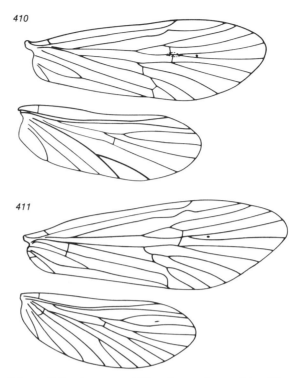

Figs. 410–411: Wings of *Brachycentrus* (*Oligoplectrum*) *maculatus* (Fourcroy, 1785)
410. male; 411. female

the simple, unisegmented inferior appendages are obliquely directed upwards and backwards, basally tightly adpressed to segment IX, with their free distal part slightly broadened and bent downwards, apex obtuse; the ventral shape of the distal parts of the gonopod is unmistakable: a deep sinus of the median border is distally followed by a strong, round swelling, and the upper median angle is sharply pointed and directed mediad. Phallus a simple tube, deeply split on the ventral side at its apex (it is represented in Fig. 414); inside the endotheca, a very distinct, large, U (or V) shaped sclerite ("phallothremmal sclerite").

Female genitalia: The abdominal segments are highly telescopic; in the female after oviposition, the abdomen is markedly shrunk, with a distinct apical cavity. The female is frequently caught with the greenish and conspicuous egg-mass retained in this apical cavity (pouch). Segment VII very slightly modified, tergite VIII large, tergite IX much smaller, both with pair of apodemes; below these two tergites only one enormous sternite (VIII? or VIII+IX?) with sclerotized, darker lateral borders, and readily becoming concave (this peculiar sternite plays an important role in the formation of the apical pouch already described); segment X small, but rather complex, concave ventrally; spermatheca strongly sclerotized.

Remarks: One character clearly distinguishes the Levantine specimens from the European ones — the medio-distal plate on sternite VII of the male is elongated,

198

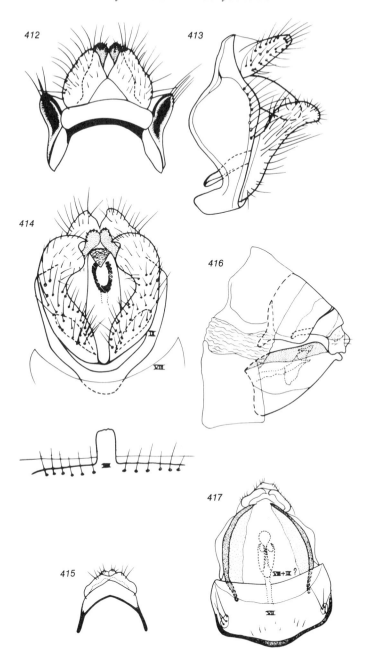

Figs. 412–417: *Brachycentrus* (*Oligoplectrum*) *maculatus* (Fourcroy, 1785)
412–414. male genitalia
(412. dorsal view; 413. lateral view; 414. ventral view);
415–417. female genitalia
(415. segments IX + X, dorsal view; 416. the same, lateral view; 417. the same, ventral view)

199

moderately slender, and smooth, whereas it is very much broader, slightly shorter, covered with minute tubercles, at least in specimens from Central Europe.

Distribution: This species has a wide European distribution, from Spain to the Caucasus, being known also from Asia Minor, but it is absent from several parts of the Continent (e.g., Italy, Greece, the British Isles, Fennoscandia). It is beyond doubt that it has in Lebanon the southernmost outposts of its distribution; it was regularly caught there at only one locality: the Orontes at Hermel, downstream from its main spring, Zarka, at 650 m a.s.l. It will be certainly found in northern Lebanon and in Syria.

Ecological Notes: In most of its distribution area, *Brachycentrus* (*O.*) *maculatus* is an inhabitant of larger watercourses (Hyporhythral, Epipotamal) with stony substrates and current of varying speeds. It is possible that the unique known Levantine locality, a river some 10–15 m broad, belongs to the Hyporhithral, too. In several parts of Europe the populations of this species are enormous. This is certainly not the case in the Levant: although it was caught regularly on the Orontes at Hermel, only comparatively few specimens were taken there — at the periphery of the distribution area. These were caught either in May, or from August to October.

Family LIMNEPHILIDAE Kolenati, 1848

Gen. Spec. Trich., 1:30, 35 (as Limnophiloidea)

Type Genus: *Limnephilus* Leach, 1815.

Medium-sized or large insects, showing great diversity in habitus and colour. Ocelli present. Maxillary palpi: 3 segments in male (1st very short, 2nd and 3rd long and subequal), 5 segments in female; segments slender, normal, not setose. Antennae moderately strong, at most as long as forewing, scapus not longer than head. Legs often with mostly dark spines on tibiae and tarsi; various formulae of spurs, but never more than one spur on forelegs and more than 3 on intermediate legs (most common formula: 1, 3, 4). Wings often parchment-like and with scanty pubescence (sometimes densely clothed). Forewings clearly differing from hindwings: forewings narrow at base, broadened at level with the anastomose, apex often elliptical; hindwings shorter, much broader, less pubescent, anal area ample, folded. Venation constant and rather complete; f 1, 2, 3, 5 in both wings; in forewing discoidal and thyridial cells long and closed, no median cell; in hindwing discoidal cell closed or open. Male genitalia extremely varied in different subfamilies; in Limnephilinae, superior, intermediate and inferior appendages always easily recognized, all rooted more or less in the bottom of an apical cavity, gonopods unisegmented and immobile; in Apataniinae all appendages slender, elongate, with tendency to longitudinal splitting, gonopods bisegmented and mobile. Female genitalia less varied, a "vulvar scale" generally present, often distinct, trilobed.

Key to the Genera of Limnephilidae in the Levant

1. Delicate insects. In forewing R 1 terminally united to C by a cross-vein into which SC runs. Two spurs on intermediate tibia.   **Apatania** Kolenati
– More robust and large insects. Forewing always lacks this peculiarity. Three spurs on intermediate tibia (the exception being *Limnephilus turanus hermonianus*: 2 spurs)   2
2. Forewings apically obliquely truncate. In hindwings, apical cell (not fork !) IV basally limited by straight cross-vein, much shorter than that of apical cell II.
      Limnephilus Leach
– Forewings apically parabolic. In hindwings, apical cell IV basally limited by oblique cross-vein longer than that of apical cell II   3
3. Male genitalia with strongly sclerotized and deeply excised superior appendages (not in all species!); female genitalia with bifid median lobe of "vulvar scale".
      Mesophylax McLachlan
– Male genitalia with normally sclerotized, not deeply excised superior appendages; female genitalia with entire median lobe of "vulvar scale".   **Stenophylax** Kolenati

Genus APATANIA Kolenati, 1847
Allg. dt. naturht. Ztg, 2(5–6): folio page appendix

Type Species, by monotypy: *Apatania vallengreni* (sic! correct name: *wallengreni*) McLachlan, 1871 (this being a *nomen novum* for *Phryganea vestita* Kolenati, 1847 nec Zetterstedt, 1840). See discussions in Kimmins (1959) and Schmid (1953:148).
Diagnosis: Slender, delicate insects, small to medium sized, body dark brown or black. Antennae shorter than forewings. Legs slender, spurs 1, 2, 4 (in some species 1, 2, 2). Wings uniformly coloured, with fine setae not obscuring venation; forewings elongated, elliptical, hindwings slightly broader with postcostal border emarginated at level of f5; both wings with f1, 2, 3, 5. Forewing with a most characteristic feature in the venation: R 1 united terminally with C by a long cross-vein, with SC running into this cross-vein; pterostygmal area conspicuous; anterior border of discoidal cell concave. Hindwing with SC and R 1 running very close to one another; discoidal cell open; f1 short; M furcates at about the same level as Cu; M 3 + 4 in its basal part touching the basal part of Cu1, connected to it by a cross-vein, or even having a short section in common with it. Male genitalia with all parts elongated, more or less slender; segment IX a sclerotized ring; segment X extremely varied structurally, a general description being impossible (it is possible in many cases to recognize a dorsal complex of generally paired, slender appendages: "internal and external branches", these latter branches often with roughly triangular shape, in lateral view); in some species — not the Levantine one — there are also small, ovoidal "superior appendages", also belonging to segment X; inferior appendages long, always bisegmented, articulated with segment IX, and thus mobile; phallic complex, when invaginated, contained in a sack (the phallocrypt), when devaginated allowing distinction of a basal phallotheca, a membranous endotheca, distally followed by an

upper pair of titillators and a lower phallus (aedeagus?). In the female genitalia, segments IX and X never with long appendages; segment IX resembling an arch, open ventrally, with the two ventral extremities forming two large, hairy lobes on both sides of the genital opening; "vulvar scale" not trilobed, but with only one, membranous "vulvar lobe"; segment X variously shaped, sometimes well protruding and more slender than segment IX, ventrally with the large anal opening; extremely large and complex "vaginal apparatus".

General Remarks: *Apatania* is a moderately large genus, with some 50 species presently known. The genus is very heterogeneous and was divided, in the monographic revision by Schmid (1953, 1954), into various supergroups, groups and subgroups of species. All the species are inhabitants of the Northern Hemisphere; 15 of them are Nearctic, the remaining being known from Eurasia (most of these are Palaearctic, quite a few Oriental). Most *Apatania* are septentrional elements, sometimes extending further north than any other caddisfly; many species are endemic to some mountain ranges, some species being typical glacial relicts. The southernmost European species is the Levantine *A. cypria*. Most species are psychrostenothermous inhabitants of springs and of the smallest streams. The larval and pupal cases are built from sand, conical in shape, but slightly flattened dorso-ventrally, short, curved, with anterior opening much wider than posterior opening.

Apatania cypria Tjeder, 1951

Figs. 418-425

Apatania cypria Tjeder, 1951, *Commentat. biol.*, 13(7):3–5. Schmid, 1954, *Tijdschr. Ent.*, 97(1/2):8–9.

A slender caddisfly. In a sample of 215 specimens from Nahal el Barouk (Lebanon), the wing expanse of the ♂♂ ranged from 11.5 to 17.6 mm, that of the ♀♀ from 13.2 to 17.3 mm (the ♂ holotype from Cyprus was described as having forewings 9 mm long, which would give a wing expanse of some 19 mm). Head, meso- and metathorax very dark brown (blackish), warts slightly paler; prothorax and abdomen brown; antennae dark brown; coxae, trochanters and femora dark brown, tibiae and tarsi lighter, with black spines. Forewings (Fig. 422) uniformly brownish with fine, short setae, not obscuring venation; pterostygma quite distinct in rubbed specimens; hindwings paler. Schmid (1954:8) writes that there is strong sexual dimorphism in the wings. This is rather strange, because the female was unknown at that time; in any event, I could not detect the slightest sexual dimorphism.

Male genitalia: The terminology in Schmid (1954) is partly adopted here. Segment IX shorter dorsally and ventrally than laterally, without any lobe from the middle of the distal border of the tergite. Segment X forming a dorsal and a ventral complex. The dorsal complex includes a pair of internal and a pair of external branches, both

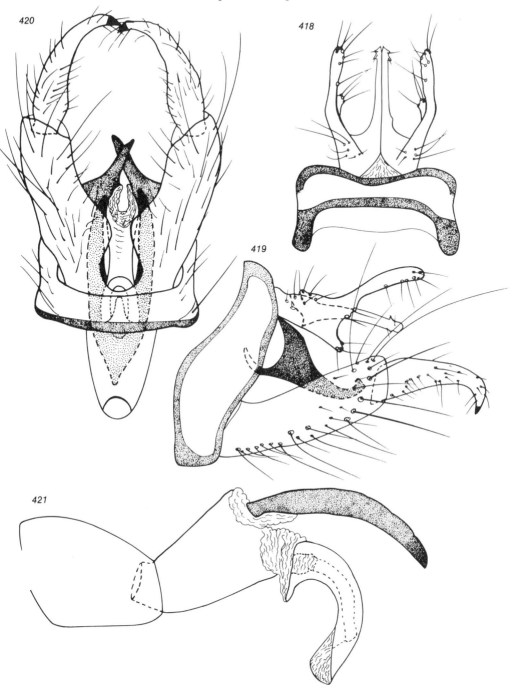

Figs. 418–421: *Apatania cypria* Tjeder, 1951, male genitalia
418. dorsal view (segment IX and dorsal complex of segment X);
419. lateral view; 420. ventral view;
421. phallic apparatus, lateral view, in complete extension

203

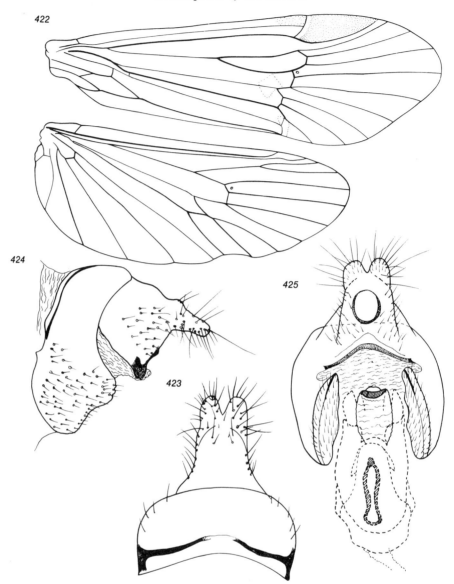

Figs. 422–425: *Apatania cypria* Tjeder, 1951
422. wings;
423–425. female genitalia
(423. dorsal view; 424. lateral view; 425. ventral view)

elongated; internal branches straight, slender, tightly adpressed (only exceptionally crossing each other, as in Tjeder's figure), apically with a pair of minute lateral points; external branches moderately broad at base, then characteristically forked, lower branching much shorter than upper branching which is slender and obliquely directed upwards and backwards. The ventral part ("body") of segment X consisting of one pair of strong, flat appendages, broad at base, then suddenly slender, terminal part bent upwards, ending in a sharp point directed proximad. Inferior appendages with long, quadrangular coxopodite, and much shorter, slender harpago with very sharply pointed tip directed mediad and downwards. Phallic complex having, as distal parts: (a) the phallus (also interpreted as aedeagus), which is a strong tube, rather short, strongly curved downwards, membranous apex deeply cleft (Figs. 420, 421); (b) a pair of lateral titillators, which are very strong, and longer than the phallus (nevertheless, not reaching its apex, because they are rooted more proximally).

Female genitalia (the female was never described): Segment IX without medio-dorsal lobe; its setose ventral parts are very strongly developed, this being evident in lateral as well as in ventral view. Setose segment X, massive laterally but moderately slender dorso-ventrally, divided into a dorsal, strongly protruding part which is deeply cleft medio-distally, and a ventral, much less protruding lobe; in lateral view (Fig. 424) dorsal and ventral lobe are separated by an oblique line with a small accident above the ventral lobe. The membranous "vulvar lobe" (seen in Fig. 425 between the ventral lobes of segment IX) has convex margins and is slightly depressed apically. Large and complex "vaginal apparatus".

Remarks: There are frequent anomalies in the venation (e.g., presence of additional, aberrant cross-veins). The structure of the male genitalia is almost identical to that of *A. cypria* as described from Cyprus (the apparent difference in size was already noted). According to Schmid (1954), *A. cypria* belongs to the *wallengreni*-group, which includes the "most typical" *Apatania*; it is very closely related to a Caucasian species, *A. subtilis* Martynov, 1909, and to a species from Asia Minor, *A. olympica* Malicky, 1982, forming with these sister-species the *subtilis* subgroup.

Distribution and Ecological Notes: This southernmost European species of *Apatania* was described from Cyprus (Troodos). In the Levant it was caught in the catchment of the Orontes, as well as in those of two small coastal rivers, Nahr ed Damour and Nahr el Aouali. It is certainly absent from Israel. In the Orontes catchment it is presently known from various springs and spring-brooks of Yammouné, at Chlifa, and at Baalbek, between 1,100 and 1,400 m a.s.l.; in the two mentioned coastal basins, the known localities are some main springs of two rivers (Nabaa es Safa, Nabaa el Barouk) and the streamlets of these springs, between 950 and 1,080 m a.s.l. We have here one of the rare typical cases of a Levantine crenobiont of higher mountains. It is very interesting to note that, whereas considerable populations are present in large springs with a stony substrate, as well as in their streamlets (Eucrenal, Hypocrenal) — a good example being Nabaa el Barouk where considerable concentrations were observed, *A. cypria* is only quite sporadically represented slightly more downstream (in the Epirhithral, probably). The species is most certainly psychrostenothermous, the optimal water temperatures probably being between 6 and ca. 12° C. It was observed on the wing throughout the year.

Genus LIMNEPHILUS Leach, 1815

In: Brewster, *Edinburgh Encyclopedia,* 9, 1 Entomology, 1815:136

Type Species: *Phryganea rhombica* Linnaeus, 1758, by monotypy.

Note: In numerous publications, starting from 1839, the generic name *Limnophilus* was used.

Diagnosis: Small to large species, with extremely varied wing patterns and genital structures. Two characters of the wings will almost invariably help one to recognize a species of *Limnephilus*: the generally parchment-like, slightly pubescent, elongate forewings are obliquely truncate at their apex, and in the much broader hindwings the apical cell (not fork!) IV is basally limited by a very short, straight cross-vein, much shorter than that of apical cell II. Spurs, in both sexes of most species: 1, 3, 4. In the middle of abdominal sternites VI and VII in the male, and V and VI in the female, there is often a small "tooth" (or point) on a false suture. Male genitalia with distinct superior, intermediate, and unisegmented, mostly small inferior appendages; phallic apparatus mostly with conspicuous, apically furcate titillators, the two branches with spines or setae.

General Remarks: This is a very large genus with some 160–180 known species ("some 200" according to Schmid, 1980:109). The species are either Palaearctic or Nearctic, a few of them being Holarctic; the northern zones of the Northern Hemisphere, especially those of the subarctic, have a distinctly richer fauna of *Limnephilus* than the meridional zones. Most of the species are inhabitants either of very different types of stagnant waters, or of not very fast flowing waters with rich macrophytic vegetation. The distribution areas are generally large; most species are found at low altitudes, but there are also mountain species. The larvae or pupae of all *Limnephilus* species build tubular cases, but there is a vast interspecific variability in their size, shape, and building materials used; moreover, there is often an important intraspecific variability. An attempt to divide the genus into a number of monospecific or polyspecific species-groups was made by Schmid (1955:133–144); he considers these groups as more or less natural, although their relationships are completely obscure, and also remarks that the number of isolated species is very large, rendering impossible a "clear and commodious classification".

Only two extremely different species are known with certainty from the Levant, but a few others are undoubtedly present in Lebanon and in adjacent areas of Syria (probably not in Israel). There is already good but incomplete evidence, from larval and pupal material, that two other *Limnephilus* species coexist with *L. lunatus* in the localities Yammouné and Chlifa. The description of *L. turanus hermonianus* nov. ssp. will be found in this volume under "Addenda" (p. 276). I consider a key to the two presently known *Limnephilus* species from the Levant as being superfluous.

Limnephilus lunatus Curtis, 1834

Figs. 426–433

Limnephilus lunatus Curtis, 1834, *Phil. Mag.*, 4:123. Neboiss, 1963, *Beitr. Ent.*, 13(5/6):586–587, 607–608.
Limnophilus lunatus auct. (e.g., McLachlan, 1875, *A Monographic Revision & Synopsis*, pp. 61–63; Schmid, 1955, *Contr. étude Limnephilidae*, pp. 133, 136).

A few ♂ specimens from Lebanon have a wing expanse ranging from 31 to 35 mm: one of the largest caddisflies in the Levant. A pale insect, most parts of body are yellow or beige; on tibiae and tarsi numerous black spines; the forewing has a variable pattern, as in Fig. 426, the dark patches being beige, pterostygma still darker, the large, light, apical semi-lunar patch characteristic for many populations of this species being absent in the available Lebanese specimens; hindwing hyaline (excepting pterostygma), with a "beard" — i.e., a short row of rigid black setae — on R 2. In the male, abdominal sternites VI and VII with one sharp point in the middle; in the female, sternite V with one minute tooth, and sternite VI with one obtuse, larger tooth. Male genitalia: Tergite VIII medio-distally with swollen and slightly protruding cushion covered with black spines. Segment IX laterally moderately well developed, less so ventrally and especially dorsally. The largest of all the appendages are the superior ones; they are rounded, posteriorly concave, with margins characteristically thickened and blackened, upper median angle produced in a conspicuous black tooth; in lateral view quadrangular, in dorsal view posteriorly emarginate. Intermediate

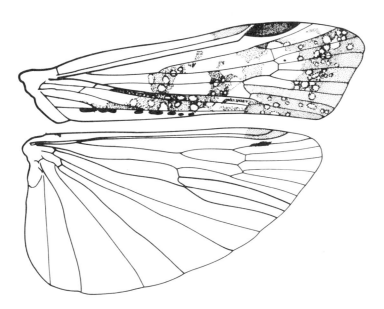

Fig. 426: *Limnephilus lunatus* Curtis, 1834, male wings

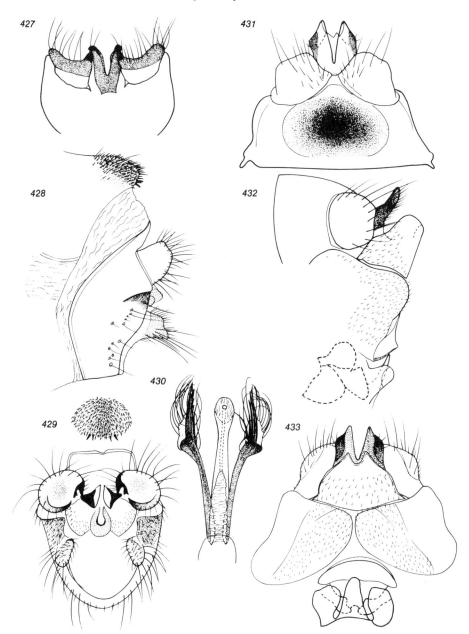

Figs. 427–433: *Limnephilus lunatus* Curtis, 1834
427–430. male genitalia
(427. superior and intermediate appendages, dorsal view;
428. lateral view; 429. posterior view;
430. distal parts of phallic apparatus, dorsal view);
431–433. female genitalia
(431. dorsal view; 432. lateral view; 433. ventral view)

appendages small, dorsally placed, not protruding beyond superior appendages in lateral view, black, roughly triangular, strongly divergent. Inferior appendages very small, only slightly protruding; their quite narrow inferior parts, meeting on the median line to form a semicircle, closely line the distal limit of segment IX; their free superior part shorter than the superior appendages, angular with distal limit angularly excised in lateral view. Phallic apparatus with very conspicuous titillators which are apically branched, dorsal branch short, strongly sclerotized, with long, strong, black spines, ventral branch longer and broader, paler, with numerous finer setae; phallus tubular, slightly capitate.

Female genitalia: Tergite IX with important concavity; sternite IX split on the median line into two large plates. Two large, more or less quadrangular plates with distal margin oblique in dorsal view, may merit being termed superior appendages. The most strongly protruding and sclerotized appendages are the intermediate ones; they are partly blackened; and the apical margin of each appendage is oblique and slightly emarginated, inner distal angle pointed. Basally these appendages are coalescent with the dorsal parts of a large ventral piece situated below them; this piece is convex, with a rounded medio-distal sinus. Vulvar scale (i.e., most proximal part of the external genitalia in ventral view) trilobed, median lobe longer and more slender than lateral lobes.

Remarks: *L. lunatus* belongs to a small, imperfectly defined group of European species; it is very closely related to *L. minos* Malicky, 1970, from Crete. It is an extremely variable species, especially with regard to the pattern of the forewings.

Distribution and Ecological Notes: *L. lunatus* has a vast distribution area, including all of Europe, North Africa, Asia Minor and northern Iran. In the Levant, it is presently known only from a few localities in the karst zone between the upper basins of the Orontes and the Litani: Yammouné, Chlifa, and Baalbek (1,100–1,400 m a.s.l.). It will certainly be found in other Levantine localities, but is probably absent from Israel. In the three mentioned localities it inhabits karst springs and their streams, thus water courses with low, constant temperature. This is clearly a case of ecological vicariance, the species being elsewhere mostly an inhabitant of stagnant water, or of slowly running water courses with abundant vegetation. The few known specimens were caught in October.

Genus STENOPHYLAX Kolenati, 1848
Gen. Spec. Trich., 1848, 1:32, 62

Type Species: *Stenophylax striatus* Kolenati nec L., a synonym of *permistus* McLachlan (subsequent selection by Kimmins, 1950).

Diagnosis: The species of *Stenophylax* (Fig. 434) rank among the largest caddisflies in the Levant. Habitus of all species very similar; parchment-like forewings very shortly setose and often characteristically and rather uniformly coloured: reddish-brown or light brick red, sometimes with faint greyish or orange tint, with very

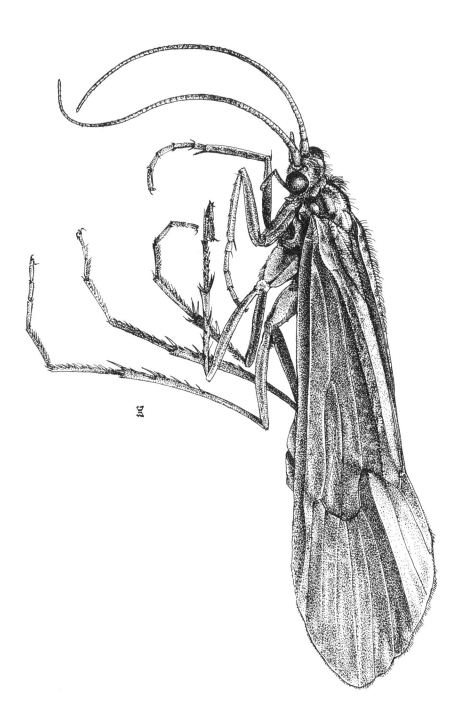

Fig. 434: *Stenophylax malaspina* (Schmid, 1957)

faint pattern of small, round spots. Forewings of elegant form, apex regularly parabolic; hindwings much broader, the anal field being strongly developed; pattern of venation typical for the family, upper border of discoidal cell in forewing depressed (concave); forks 1, 2, 3, 5 in both wings. Spurs in some species 0, 3, 4 in the male and 1, 3, 4 in the female, in other species 1, 3, 4 in both sexes. First tarsal segment of male forelegs in some species longer than or as long as 2nd segment, in other species distinctly shorter than it. Male genitalia always with the following characters: tergite VIII with a zone covered by dark spinules (tubercles), sometimes bipartite on median line; segment IX more strongly developed laterally, less so ventrally and especially dorsally; distinct superior, intermediate and inferior appendages; superior appendages ear-shaped or foliaceous; intermediate appendages strongly sclerotized, dark, arising from strongly sclerotized plates which surround a space with the anal opening; below anus, a subanal plate, often inconspicuous; inferior appendages unisegmented, well developed in a vertical plane, laterally clearly or indistinctly delimited from segment IX, with apical free part slightly or strongly protruding and varying considerably in shape in various species; phallic complex with phallus and a pair of strongly sclerotized titillators. Female genitalia: dorsal part of segment IX forming a collar, ventral part forming two setose, swollen lobes, more or less widely separated on the median line; dorsal part of segment X represented by a pair of appendages or lobes (sometimes called "superior appendages"), in some species with a third, median lobe; just below these lobes, in some species only, a well sclerotized plate interpreted as "ventral plate (or scale) of segment X"; internal vaginal apparatus complex; the most proximal position in the genitalia is occupied by a distinctly trilobed "vulvar scale", and just above it there is a large, hemi-ellipsoidal "supragenital plate".

General Remarks: It has long been known that *Micropterna* Stein, 1874, cannot be considered as a genus really distinct from *Stenophylax*, because there is not a single character enabling separation of these taxa in all cases. Some recent authors have persevered in considering them as distinct, merely for traditional reasons, whereas others implicitly abandoned *Micropterna*. I am doing this explicitly, considering *Micropterna* as a synonym of *Stenophylax*. *Stenophylax*, in the present sense, is an interesting genus including about 30 known species (for a revision see: Schmid, 1957; but several species were described later, and there are problems concerning the validity of some of them). Several small groups of species can be recognized, one of them — probably the largest — being represented by the species of "*Stenophylax* sensu stricto". Most *Stenophylax* are inhabitants of parts of the Periponto-Caspi-Mediterranean zone, and especially of the Near East where the largest number of species is present. A few species have penetrated into Central and Northern Europe, but only a very small number are found outside the limits of the western Palaearctic: one in Afghanistan, one in India and one in China (Szechwan). The species of *Stenophylax* are ecologically, ethologically and phenologically a very distinctive group; larvae and pupae develop mainly in springs and streams, often of karstic or semi-arid areas; larval cases are large, cylindrical or slightly conical, built from rough sand, pupal cases are vertically thrust into the substrate; species strictly monovoltine, imaginal emergence during spring; after emergence the adults (generally excellent flyers) migrate to subterranean hollows (caves, etc., not very far from the entrances; they are "subtroglophile" insects) or to cracks in the rocks, for a long summer period of quiescence during which mating and maturation of eggs occur; during last months of the year, at least the (gravid) females abandon their subterranean shelters for oviposition, the development being restricted to the winter months.

On some more species possibly present in the Levant

The distributional peculiarities of the group, and the fact that it was very unsatisfactorily searched for in Lebanon, and not at all in limitroph zones of Syria, or in Jordan, make it not unlikely that species other than the 6 described here will be discovered in the Levant. This is more plausible for the following species:

(A) *S. caesareicus* (Schmid, 1959). Described as *Micropterna*. Not to be confused with *Stenophylax caesareus* Navas, 1917 (a nomen nudum). *Micropterna ariadne* Malicky, 1970, is a synonym. *S. caesareicus* (Schmid) was mentioned from Asia Minor, the European part of Turkey, Greece, Crete, Bulgaria, and Yugoslav Macedonia. The drawings of genitalia (Figs. 435–443) are reproduced from Schmid (1959) and from Malicky (1985).

(B) *S. muehleni* (McLachlan, 1884). Described as *Micropterna*. Mentioned from the Caucasus, the Caspian zone, Turkmenia, Asia Minor, Iran and Afghanistan. The drawings of genitalia (Figs. 444–449) are reproduced from Schmid (1957) and from Malicky (1983).

(C) *S. solotarewi* (Martynov, 1913). Described as *Micropterna*. *Micropterna caspica* Schmid, 1959, is a synonym. Known from the Caucasus and from northern Iran. The drawings of genitalia (Figs. 450–454) are reproduced from Schmid (1959).

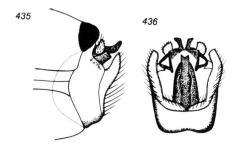

Figs. 435–436: *Stenophylax caesareicus* (Schmid, 1959), male genitalia
(435. lateral view; 436. posterior view.
From Schmidt, 1959)

212

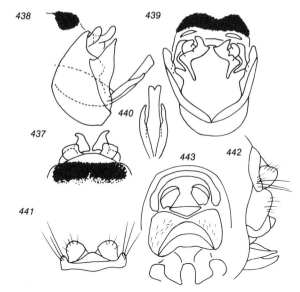

Figs. 437–443: *Stenophylax caesareicus* (Schmid, 1959)
437–440. male genitalia
(437. dorsal view; 438. lateral view; 439 posterior view; 440. phallic apparatus);
441–443. female genitalia;
(441. dorsal view; 442. lateral view; 443. posterior view).
(From Malicky, 1985)

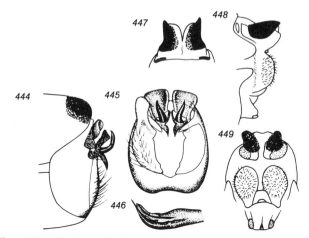

Figs. 444–449: *Stenophylax muehleni* (McLachlan, 1884)
444–446. male genitalia
(444. lateral view; 445. posterior view;
446. apex of phallic apparatus, lateral view.
From Schmid, 1957);
447–449. female genitalia
(447. dorsal view; 448. lateral view; 449. posterior view.
From Malicky, 1983)

213

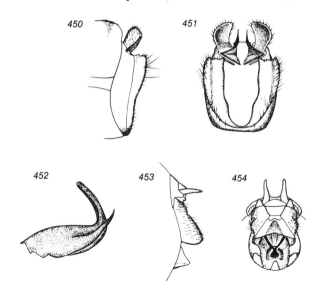

Figs. 450–454: *Stenophylax solotarewi* (Martynov, 1913)
450–452. male genitalia
(450. lateral view; 451. posterior view; 453. phallic apparatus, lateral view);
453–454. female genitalia
(453. lateral view; 454. posterior view).
(From Schmid, 1959)

Key to the Species of Stenophylax presently known from the Levant

♂♂

1.	First tarsal segment of forelegs longer than second	2
–	First tarsal segment of forelegs shorter than second	4

2. Superior appendages (dorsal view) claw shaped; intermediate appendages (viewed from behind) parallel; inferior appendages (lateral view) only with very short free apical part. Spurs 1, 3, 4. **S. permistus** McLachlan

– Superior appendages not claw shaped; intermediate appendages distinctly divergent; inferior appendages with important free apical part 3

3. Inferior appendages (lateral view) with slender, stripe shaped, free distal part. Spurs 1, 3, 4. **S. vibex** (Curtis)

– Inferior appendages with strong, conical, free distal part. Spurs 0, 3, 4.
S. tauricus Schmid

4. Phallus strongly furcate at end. Blackened spiny zone of tergite VIII broad like the tergite itself but short, only medially roundly produced **S. malaspina** (Schmid)

– Phallus not strongly furcate at end. Blackened spiny zone of tergite VIII with different shape 5

5. Inferior appendages laterally only slightly protruding, viewed from behind very broad, with medio-apical angles produced in short but strong blackened "shoulders". Superior appendages, small, simple. **S. coiffaiti** (Décamps)

214

−   Inferior appendages laterally very strongly protruding, conical, viewed from behind very slender, medio-apical angles only slightly produced. Superior appendages large, with complex relief, viewed from behind with medio-apical angles distinctly protruding mediad.   **S. lindbergi** (Tjeder)

♀♀
(♀ of *S. tauricus* unknown)

1.   Lateral lobe of vulvar scale (lateral view) with distinct incision of upper margin; lobes of segment X dorsally and ventrally separated by a kind of median lobe, but this belongs to segment IX (not X).   **S. coiffaiti** (Décamps)
−   Lateral lobe of vulvar scale without incision of upper margin; lobes of segment X either without median lobe between them, or with a median lobe also belonging to segment X
                                                                                          2
2.   Dorsal part of segment X with two lateral lobes and one median lobe   3
−   Dorsal part of segment X without median lobe (in *S. malaspina* with very faint median lobe, belonging to segment IX, not X)   4
3.   Lateral lobes of segment X very small, only indistinctly protruding beyond distally emarginate, large median lobe. Ventral plate of segment X rather narrow, not reaching its lateral margins. Dorsal part of segment IX (lateral view) protruding backwards at upper and lower end.   **S. permistus** McLachlan
−   Lateral lobes of segment X distinctly protruding beyond trapezoidal median lobe, and with small median projection near base. Ventral plate of segment X broad, reaching its lateral margins. Dorsal part of segment IX (lateral view) not protruding backwards at upper and lower end.   **S. vibex** (Curtis)
4.   Very conspicuous ventral plate ("scale") of segment X. Ventral lobes of segment IX (viewed from behind) with wide semicircular space in between. Lobes of segment X apically pointed (lateral view).   **S. lindbergi** (Tjeder)
−   No conspicuous ventral plate of segment X. Ventral lobes of segment IX only with small space in between. Lobes of segment X apically obtuse (lateral view).
                                                                    S. malaspina (Schmid)

Stenophylax vibex (Curtis, 1834)
Figs. 455–463

Limnephilus vibex Curtis, 1834, *Phil. Mag.*, 4:125.
Stenophylax vibex —. McLachlan, 1875, *A Monographic Revision & Synopsis*, p. 136, Pl. 14, figs. 1–5; Schmid, 1957, *Trab. Mus. Cienc. nat. Barcelona*, N. ser. zool., 2(2):10–11; Neboiss, 1963, *Beitr. Ent.*, 13(5/6):627–628, etc.; Malicky, 1980b, *Entomofauna* (Linz), 1(8):97–98, Figs.
Stenophylax speluncarum McLachlan, 1875, *A Monographic Revision & Synopsis*, p. 136, Pl. 14, figs. 1–4.
Stenophylax vibex vibex —. Schmid, 1957, *Trab. Mus. Cienc. nat. Barcelona*, N. ser. zool., 2(2):11–12; Schmid, 1959, *Beitr. Ent.*, 9(7/8):784.
Stenophylax vibex speluncarum —. Schmid, 1957, *Trab. Mus. Cienc. nat. Barcelona*, N. ser. zool., 2(2):12–13.

Stenophylax meridiorientalis Malicky, 1980b, *Entomofauna* (Linz), 1(8):98, Figs.
Stenophylax minoicus Malicky, 1980b, *Entomofauna* (Linz), 1(8):98–99, Figs.
Stenophylax zarathustra Malicky, 1980b, *Entomofauna* (Linz), 1(8):99, Figs.

A very large species, one of the largest in the Levant, wing expanse from slightly less than 40 mm to more than 50 mm; in the unique ♂ from Lebanon which I could examine: 47.5 mm. Spurs: 1, 3, 4. First segment of tarsus of forelegs much longer than 2nd segment.

Male genitalia (description based on the Lebanese specimen): Tergite VIII with two roughly triangular zones covered with black spinules, almost joining on median line. Superior appendages elongate, simply ear-shaped in lateral view, but strongly concave medially, viewed from behind with a distinct but not very large projection on the inferior border in its proximal part. Intermediate appendages very strong, blackened in their distal parts, in lateral view obliquely directed backwards and upwards to the pointed and slightly curved tip, viewed from behind very strongly geniculated in their middle, distal parts very strongly diverging, each appendage roughly describing a semicircle; the sclerotized plates from which the intermediate appendages arise are moderately broad but with slender roots. Inferior appendages in their proximal parts tightly adpressed to segment IX, nevertheless with a significant free part; in lateral view very narrow, free part strongly diverging from segment IX, apically truncate and with very small excision; viewed from behind they are distinctly broader than in lateral view, near the base of the internal border with small emargination, distal border emarginate, medio-apical angle distinctly and bluntly produced. Titillators curved, with a very small number of minute teeth on distal part of their external border.

Female genitalia (description based especially on Schmid, 1957, the drawings here reproduced being made from Western European specimens representing "*S. vibex vibex*" and from Eastern European specimens representing "*S. vibex speluncarum*": Figs. 458–460 and 461–463 respectively): Despite the variability, the dorsal shape of segment X unmistakably characterizes this species: median lobe very well separated from lateral lobes, median lobe always shorter, trapezoidal, lateral lobes conical, more or less slender, always with distinct but small projection on basal part of their median margin. Other characters: dorsal part of segment IX, in lateral view, without superior and inferior lobes; ventral plate of segment X always trapezoidal, varying in length and width.

Remarks: *Stenophylax vibex* is a species with variable male and female genitalia, but its variability has been interpreted in different ways. Malicky (1980b) describes three new species from this complex, one of them being *S. meridiorientalis* (as a matter of fact a new name for *S. speluncarum* McL., this last name being considered by the author as invalid because the male lectotype of *S. speluncarum* was found not to be distinct from British specimens of *S. vibex*). I disagree with some other conclusions of the mentioned publication. First of all, the author ignored several sources of information on the variability of *S. vibex* (e.g., Botosaneanu, 1959:46–48, general

216

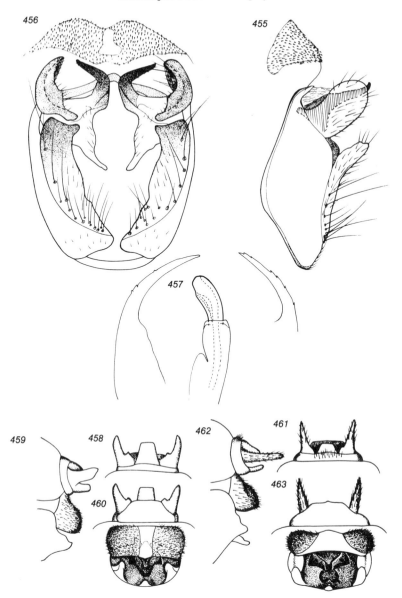

Figs. 455–463: *Stenophylax vibex* (Curtis, 1834)
455–457. male genitalia, specimen from Lebanon
(455. lateral view; 456. posterior view;
457. apex of phallic apparatus, lateral view, and apex of titillators);
458–463. female genitalia of European specimens
determined by F. Schmid as *S. vibex vibex* (458–460)
or as *S. vibex speluncarum* (461–463).
(458–463: from Schmid, 1957)

discussion on the variability; Schmid, 1959:784, remarks on specimens from Iran); even the information in Schmid (1957:10–13) was incompletely used. Further, I consider the argumentation in this paper unsatisfactory and unconvincing. The variability of this species is apparently interesting, in some cases it may have a geographic background; a careful study of this variability has yet to be performed; but *one species* is involved here, not several.

Turning now to the Levantine form, this should be *S. meridiorientalis* according to distributional data in Malicky (1980b), whereas in Malicky & Sipahiler (1984:211) the Levant is directly mentioned. But male genitalia of the unique Lebanese specimen which I was able to study, clearly show that it is quite distinct from the form described as *S. meridiorientalis*. This is especially evident in the shape of the intermediate appendages and in the extremely inconspicuous development of minute teeth on the titillators, both characters pointing to "the typical *S. vibex*". The apex of the inferior appendages in lateral view is as in specimens from Eastern Europe (this was correctly represented by Schmid, 1957: Fig. 8a), not as in the figure given by Malicky for *S. meridiorientalis*. In conclusion: we are dealing here with *S. vibex* (Curtis), and only a careful study of many specimens of both sexes from the Levant and adjacent areas will enable us to say if they belong to some geographical race, and which name has to be applied to this race.

S. vibex is related to *S. tauricus* Schmid, and not far from *S. permistus* McLachlan.

Distribution: *S. vibex* is distributed in southern and central Europe (generally not in northern Europe, although recently found in southern Norway), being known also from northern Africa, Asia Minor, northern Iran and Lebanon. From this last country, the only published specimens (Décamps, 1962:580) are from one of the vertical caves of Mechmechi, above the village of Mrouge (western slopes of the Lebanon Range, some 25 km east of Beirut, at ca. 1,500 m a.s.l.), and from the cave of Harajel, or Hrajel (western slopes of the Lebanon Range, above the village of Hrajel, on the Joûnié-Faraya road, at ca. 1,500 m a.s.l.). They were caught in May and June. Of the 3 specimens (1 male, 2 female) collected, only the male could be found, and I was able to study it (courtesy of Dr H. Décamps and Dr Z. Moubayed); the specimen is presently in the Zoological Museum, University of Amsterdam. I suspect that *S. vibex* is widely distributed, especially in karst zones of Lebanon and Syria. Adults of this species are known to be typical "subtroglophiles".

Stenophylax permistus McLachlan, 1895

Figs. 464–469

Stenophylax concentricus —. McLachlan, 1875, *A Monographic Revision & Synopsis*, pp. 114, 134–135, Pl. 14, figs. 1–8 (nec Zetterstedt).
Stenophylax permistus McLachlan, 1895, *Entomologist's mon. Mag.*, 31:139–140. Schmid, 1957, *Trab. Mus. Cienc. nat. Barcelona*, N. ser. zool., 2(2):9–10.

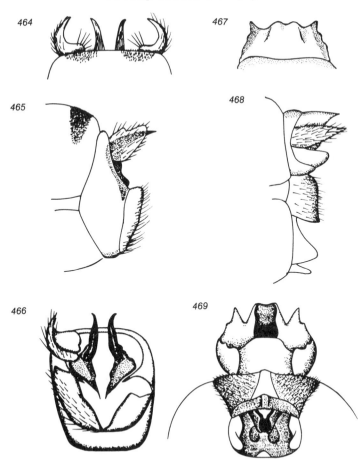

Figs. 464–469: *Stenophylax permistus* McLachlan, 1895
464–466. male genitalia
(464. dorsal view; 465. lateral view; 466. posterior view.
464: from Mosely, 1939a; 465–466: from Schmid, 1957);
467–469. female genitalia
(467. dorsal view; 468. lateral view; 469. posterior view.
467: from A. Murgoci & S. Marcoci,
Buletin sti. Acad. Republ. pop. rom., 7(2), 1955;
468–469: from Schmid, 1957)

In Malicky & Sipahiler (1984:211) this species is simply mentioned from the
"Levant"; I did not see Levantine specimens of *S. permistus*, but I include it here,
considering its presence in the Levant as probable.

A very large species, one of the largest caddisflies in the Levant, the wing expanse
ranging from about 40 mm to more than 50 mm. Spurs (♂ , ♀): 1, 3, 4. First tarsal
segment of forelegs distinctly longer than 2nd segment. Male genitalia readily

recognized especially by the dorsal shape of the superior appendages (Fig. 464) which are deeply excised medially and therefore claw-shaped (the two appendages apparently forming a "forceps"), with lateral lobe sharply pointed, curved mediad, much longer than the small, blunt median lobe; these peculiarities are evident also in lateral view, though less striking. Other characters of the male genitalia are: posteriorly on tergite VIII two patches of dark spinules, with a pale area between them; intermediate appendages very strong, almost straight, divergent only at their apices; inferior appendages only very slightly protruding, with distal margin (lateral view!) convex, very well characterized by the fact that they are proximally almost entirely welded to segment IX, only their blunt extreme apices being free. In the female genitalia, the most distinctive character is offered by segment X in dorsal view: its central part is large, trapezoidal, with emarginated apical border, whereas its lateral lobes — separated from the central part by deep, rounded or angular excisions — are triangular, apices pointed, with faint "shoulders" along the lateral margins (Fig. 467). Other distinctive characters of the female genitalia: dorsal part of segment IX, in lateral view (Fig. 468) produced in a superior and an inferior lobe; ventral plate of segment X (Fig. 469) relatively narrow, i.e., not reaching its lateral borders.

Stenophylax permistus is related — not very closely — to *S. vibex*, and was sometimes confused with it. It is not a very variable species. Known from most of Europe and from Asia Minor (not from North Africa). The adults are known to be typically "subtroglophiles".

Stenophylax tauricus Schmid, 1964
Figs. 470–471

Stenophylax tauricus Schmid, 1964, *Opusc. zool. Münch.*, 73:4–5.

In Malicky & Sipahiler (1984:211) this species is simply mentioned from Lebanon; I did not see specimens of *S. tauricus*, which is included here because its presence in the Levant is plausible. Description and illustration are entirely based on Schmid (1964).

Wing expanse (♂): 37-38 mm. Spurs in the ♂ : 0, 3, 4. First tarsal segment of forelegs "not much longer than 2nd". Male genitalia. Tergite VIII with one short and broad zone covered with black spinules. Superior appendages elongated, resembling regularly shaped, almost vertically placed and bluntly ending lobes, having on their internal face numerous minute conical tubercles. Intermediate appendages very strong, obliquely directed upwards, clearly divergent. Inferior appendages highly characteristic, large, with free part remarkably well developed; in lateral view conical, with distinct subapical depression, apex sharply pointed, well sclerotized; in ventral view the lateral depression is not distinct, and the terminal sharp point is accompanied, on both sides, by a "shoulder" (median shoulder more distally placed than lateral shoulder). Female unknown.

220

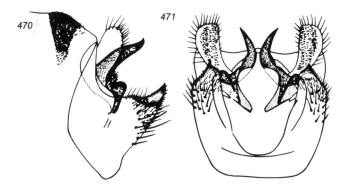

Figs. 470–471: *Stenophylax tauricus* Schmid, 1964, male genitalia
(470. lateral view; 471. posterior view).
(From Schmid, 1964)

This species is considered by Schmid as "un *Stenophylax* authentique par les caractères des genitalia du ♂ ", despite the spur formula typical for "*Micropterna*"; it is related to *S. vibex*, but clearly distinct from it. Originally described from "Taurus, Marasch"; it is certainly present in Syria, too, but has, nevertheless, a restricted distribution in the Near East. Almost nothing is known about its biology.

Stenophylax coiffaiti (Décamps, 1962)

Figs. 472–478

Micropterna coiffaiti Décamps, 1962, *Annls Spéléol.*, 17(4): 580–581.
Micropterna sp.: Schmid, 1964, *Opusc. zool. Münch.*, 73: 6–7. *Micropterna* (?) sp.: Malicky, 1970, *Ent. Z., Frankf. a. M.*, 80(14): 131–133. Type Locality: "grotte de Harajel", Lebanon. No type was mentioned in the original publication; both ♂ syntypes apparently lost (they could not be found in the H. Décamps collection at the University of Toulouse: inf. Z. Moubayed, 1984).

Wing expanse variable, ranging from ca. 30 mm to ca. 44 mm. Spurs: 0, 3, 4 in the ♂ ; 1, 3, 4 in the ♀ . First segment of tarsus of forelegs in the ♂ having slightly more than 1/2 of the length of 2nd segment.

Male genitalia: Tergite VIII with one large, laterally rounded zone densely covered with black spinules, terminally strongly bent downwards. Segment IX short in lateral view. Superior appendages very small, roughly rounded, medially concave and with fine setae. Intermediate appendages short, nevertheless protruding beyond superior appendages in lateral view, laterally roughly triangular but with sinuous dorsal border, directed upwards, viewed from behind almost touching each other on median line, pointed apically. Inferior appendages remarkably large and broad when viewed from behind, laterally very setose, only indistinctly delimited from segment IX; almost

221

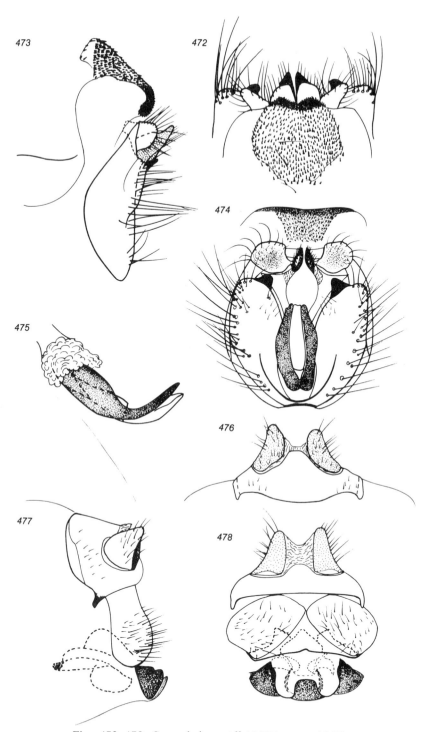

Figs. 472–478: *Stenophylax coiffaiti* (Décamps, 1962)
472–475. male genitalia (472. dorsal view; 473. lateral view;
474. posterior view; 475. phallic apparatus, lateral view);
476–478. female genitalia (476. dorsal view; 477. lateral view; 478. posterior view)

nothing protrudes from these appendages in lateral view, because they are developed perpendicularly on segment IX; their medio-apical angles protruding in characteristic, strongly sclerotized, dark, strong but short, bluntly triangular (or slightly quadrangular) points. Titillators strongly curved upwards in their distal half; phallus with almost the same shape as titillators.

Female genitalia: Dorsal part of segment IX distinctly protruding medio-dorsally as well as medio-ventrally. Segment X represented by two lateral sclerotized lobes separated by membrane; these lobes are rather small, conical, with blunt apices in dorso-ventral view. Lateral lobes of vulvar scale, especially in lateral view, with a characteristic incision of their upper margin.

Remarks: *S. coiffaiti* is related to several species formerly placed in *Micropterna*, and forming what may be termed the small group of *sequax* McLachlan, 1875.

Distribution and Ecological Remarks: The species was described from Lebanon, where the following localities are presently known: the type locality, which is the "grotte de Harajel" (in the western slopes of the Lebanon Range, valley of Nahr el Kelb, above the village of Hrajel, road from Joûnié to Faraya, ca. 1,500 m a.s.l.); Bcharré (Beshari), in northern Lebanon (ca. 1,400 m a.s.l.); and the karst spring Nabaa Mourched with its brook, in the coastal catchment of Nahr el Aouali, between El Moukhtâra and Aïn Qâniye, at 800 m a.s.l. The Lebanese specimens were caught from May to June. *S. coiffaiti* is also mentioned from Cyprus, Rhodes, and Asia Minor; a mention from "Greece" in Malicky & Sipahiler (1984:211) is probably erroneous. We are dealing here with a species having crenobiont-rhithrobiont larvae and pupae, and "subtroglophile" imagines.

Stenophylax malaspina (Schmid, 1957)

Figs. 434, 479–485

Micropterna malaspina Schmid, 1957, *Trab. Mus. Cienc. nat. Barcelona*, N. ser. zool., 2(2):34–35; Schmid, 1964, *Opusc. zool. Münch.*, 73:5; Botosaneanu, 1974a, *Trav. Inst. Spéol. Emile Racovitza*, 13:67–70.

Wing expanse in specimens from Israel: ♂ 30–37 mm, ♀ 34–42 mm. Habitus: Fig. 434. Spurs: 0, 3, 4 in the ♂, 1, 3, 4 in the ♀. First tarsal segment of forelegs in the ♂ only 1/2 of the length of 2nd segment.

Male genitalia: Posteriorly on tergite IX a zone covered by black spinules, with characteristic shape: as broad as the tergite, but very short, only postero-medially strongly protruding (downwards). Segment IX interrupted medio-dorsally; laterally segment IX coalescent with inferior appendages. Superior appendages moderately large, in lateral view more broadened distally than proximally, internally slightly concave and with fine setae. Intermediate appendages shorter than superior appendages, blackened, largely separated medially, convergent towards their tips, curiously concave on their lateral (superior) face, borders parallel, with blunt apex.

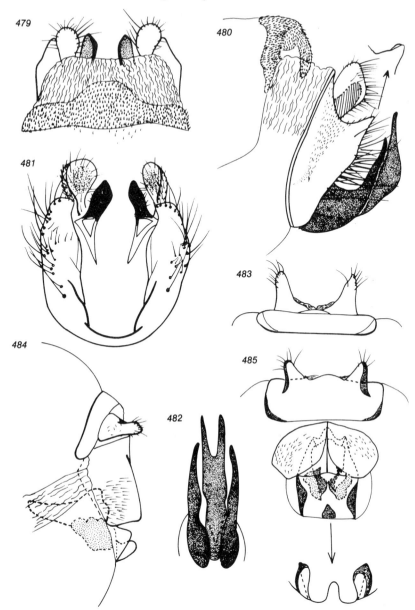

Figs. 479–485: *Stenophylax malaspina* (Schmid, 1957)

479–482. male genitalia

(479. dorsal view; 480. lateral view — arrow pointing to
figure of apex of gonopod under a slightly different angle;

481. posterior view — without phallic apparatus;

482. phallic apparatus, ventral view);

483–485. female genitalia

(483. dorsal view; 484. lateral view;

485. posterior view — arrow pointing to figure of vulvar scale, ventral view)

Inferior appendages, in lateral view, coalescent for most of their length with segment IX, but with distinct free distal part; in completely lateral view this distal part is triangular and apically pointed, but at a slightly different angle the shape of the apex may be slightly truncate; viewed from behind, the gonopods are developed on a plane perpendicular to segment IX, and are remarkably high, with median border sinuate, distally slightly capitate, medio-apical angle only very slightly protruding mediad. Phallus very characteristically furcate in its distal part; titillators shorter than it, basal part very broad, distal part much more slender and turned upwards (like phallus).

Female genitalia: Dorsal part of segment IX reduced to a rather narrow sclerotized stripe, in lateral view with superior and inferior angles not produced. Ventral part (lobes) of segment IX strongly developed, in lateral view flat and oblique posteriorly. The two (lateral) lobes of segment X are small, conical, slightly divergent, well separated on the median line. Vulvar scale with median lobe much smaller than lateral lobes.

Remarks: *S. malaspina* belongs to what could be named the *fissus* species-group, formerly placed in *Micropterna*, being related to *S. fissus* McLachlan and to *S. malatesta* Schmid. There is some variability in the male genitalia, especially in the apex of the gonopods (even within the same population, as was observed at Birket Ram); some observations on the variability of the male genitalia were published by Schmid (1964: 5).

Distribution: The species is known from Asia Minor, from Crete, from peninsular Greece, and from Bulgaria. It was caught frequently in the Golan Heights (Area 18), either on the slopes of Mount Hermon, at ca. 2,000 m a.s.l., on Mount Avital, at Naḥal Saʻar (a tributary of Naḥal Hermon), at ʻEin Qinia, in the small intermittent lake Birkat Ram, at ca. 1300 m a.s.l., as well as at Senir and in the Masaʻada forest. Other localities are known from Upper Galilee (1): it is regularly caught at Ḥuliot (Sedé Neḥemya), at only 85 m a.s.l., but it is also known from the highest mountain of Israel up to its top: Mount Meron, some 1,200 m a.s.l.; I have also seen specimens caught at Naḥal Ḥermon (Banias) and from Buteicha, some 200 m below sea level in area 7. It will undoubtedly be caught in Lebanon and in adjacent zones of Syria, too.

Ecological Notes: Observations on the biology of *S. malaspina* were published by Botosaneanu (1974a). This is one of the few species of *Stenophylax* (including *Micropterna*) whose young instars are inhabitants of stagnant water bodies (another is *S. badukus* Mey & Müller, 1979 from the Caucasus; see Mey & Müller, 1979: 178–181). A very large population was found in the small intermittent mountain lake Birket Ram, in a xeric zone of the Golan Heights; this lake sometimes dries up completely towards the end of summer; during the first days of April, shortly after the snow melted away, many "ripe" pupae were found, pointing to the early emergence of the imagines (April–May, probably during the night); subsequently, the adults obviously seek shelter for a protracted aestivation in cracks of rocks, under stones, possibly also in caves; the egg masses are laid during early winter, when the lake is again filled with water. There is good agreement between these observations and the time when adults of *S. malaspina* were caught at other Levantine localities: either in April, or in October–December; as yet, no Levantine specimen has been caught during the long aestivation period. It is quite possible that the populations in the other localities in Upper Galilee and the Golan Heights breed in similar intermittent bodies of standing water.

Stenophylax lindbergi (Tjeder, 1951).

Figs. 486–495

Micropterna taurica Martynov, 1917, *Ezheg. zool. Muz. (Petrograd)*, 21(1916):182–184.
   Schmid, 1957, *Trab. Mus. Cienc. nat. Barcelona*, N. ser. zool., 2(2):37; Décamps, 1962, *Annls Spéléol.*, 17(4):580; Malicky, 1970, *Ent. Z., Frankf. a. M*, 80(14):131, Fig. 5; Botosaneanu, 1974a, *Trav. Inst. Spéol. Emile Racovitza*, 13:65–66, Pl. II figs. d, e, f.
Micropterna lindbergi Tjeder, 1951, *Comment. Biol.*, 13(7):5
Micropterna triangularis Schmid, 1964, *Opusc. zool. Münch.*, 73:6. Botosaneanu & Gasith, 1971, *Israel J. Zool.*, 20:118.

In most of the examined Levantine specimens, wing expanse ranges, in the ♂, from 27.5 to 32.8 mm, in the ♀ from 29.5 to 36 mm; but there is considerable variability, and, e.g., in a series from Ain el Fadjar the specimens are considerably smaller: ♂ 21.5–25 mm, ♀ 26.5 mm. Spurs: 0, 3, 4 in the ♂, and 1, 3, 4 in the ♀. First tarsal segment of forelegs measuring slightly less than 1/2 of the length of 2nd segment. Wings: Fig. 486.

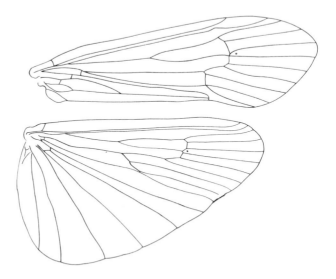

Figs. 486: *Stenophylax lindbergi* (Tjeder, 1951), wings

226

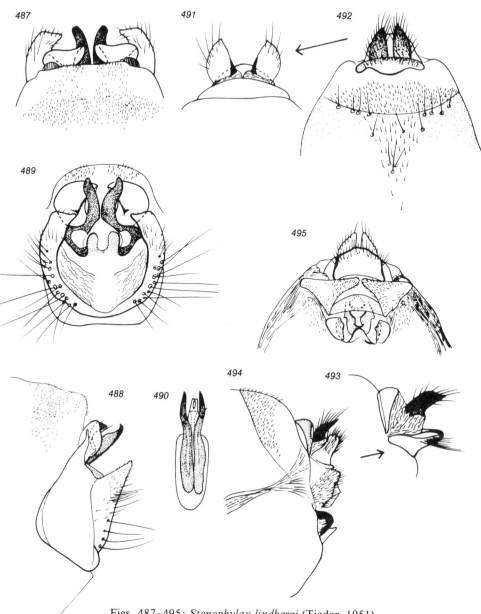

Figs. 487–495: *Stenophylax lindbergi* (Tjeder, 1951)
487–490. male genitalia
(487. dorsal view; 488. lateral view; 489. posterior view;
490. phallic apparatus, ventral view);
491–495. female genitalia
(491–492. dorsal view, two different specimens;
493–494. lateral view, two different specimens, slightly different angles;
495. posterior view).
(492, 494, 495: from Botosaneanu, 1974a)

227

Male genitalia: Posterior third of tergite VIII with dark spinules and a round medio-posterior patch devoid of them; as these spinules are very fine, rather sparse and not just black, there is no conspicuous blackened zone as in other species of the genus. Segment IX moderately well developed laterally, very narrow dorsally where it is interrupted in the middle. Superior appendages fairly large, with complex relief and very characteristic shape when viewed from behind: lateral margins rounded, median margins strongly concave, inner superior angle strongly produced, an oblique blackened keel across the median face of the appendage ending basally in a rather faint "tooth". Intermediate appendages strong, when viewed from behind resembling slightly sinuous, diverging and bluntly ending sclerotized stripes; the lateral shape of the intermediate appendage is characteristic: it ends in an upturned point, and its ventral margin is distally somehow twisted. Interestingly, the subanal plate is very large and deeply bilobed. Inferior appendages distinctly delimited laterally from segment IX; they are conical in lateral view; viewed from behind their sclerotized parts are long and narrow stripes, proximally diverging, then slightly converging (thus limiting a circular space), slightly broader at their distal end, apical borders very slightly emarginate, medio-apical angles moderately well produced. Titillators much stronger than phallus, proximally parallel but distally divergent and sinuous in a characteristic manner (shaped like a lyre), and in some specimens with extremely inconspicuous, minute tubercles.

Female genitalia: Dorsal part of segment IX slightly trilobed in dorsal view, but without protruding lobes in lateral view. Ventral part of segment IX forming two lobes, with complex relief in lateral view; viewed from behind, these lobes are roughly triangular, their median margins strongly excised and limiting a dome-shaped space. Dorsal lobes of segment X more or less triangular-foliaceous in dorso-ventral view, in lateral view obliquely directed posteriad, pointed apex very slightly upturned (their appearance varies when viewed at slightly different angles, and they are sometimes much more strongly protruding than in Figs. 493 and 494). Below these lobes is located the strongly sclerotized ventral plate of segment X, polygonal and rather short and broad in ventral view, oblique and with very obtuse apex in lateral view.

Remarks: This species was described as *Micropterna taurica*. Since *Micropterna* is now considered a synonym of *Stenophylax*, and a *Stenophylax tauricus* has already been named, a replacement name became necessary: this is the name of the oldest known synonym.

S. lindbergi is very closely related to *S. testaceus* (Gmelin, 1788) from southern and central Europe, and possibly also – but to a lesser degree — to *S. terekensis* (Martynov, 1913) from the Caucasus. The differences enabling separation of *S. lindbergi* from *S. testaceus* are small, this being true for both sexes, but there is little doubt about their being distinct species.

There is some variability in the armature of spinules of the titillators; in the original description (specimens from Crimea: Martynov, 1917, fig. 22) these are moderately well developed; in Malicky (1970, fig. 5c) a specimen from Asia Minor with long spinules on the titillators is illustrated; in the Levantine specimens there are either

228

no spinules, or these are replaced by a few very inconspicuous tubercles. In the female genitalia, the shape of the dorsal lobes of segment X may show some variability, but it is possible that we are dealing here merely with artifacts.

Distribution and Ecological Notes: *S. lindbergi* has a wide distribution, being known from Bulgaria, Crimea, Asia Minor, parts of Iran, the Levant, Cyprus, and Crete; it is certainly present in the Caucasus and in Iraq, too. We have here a typical case of geographic vicariance, the vicariant species being *S. testaceus* from southern and central Europe; the border line between the two vicariants must run through the islands of the Aegean Sea. In the Levant, the species is known from Lebanon, the Golan Heights and Israel. There is only one known Lebanese locality, and this is also the single Levantine locality where *S. lindbergi* was caught in a cave: one of the vertical caves at Mechmechi, in the western slopes of the Mount Liban (village Mrouge, some 25 km as the crow flies east from Beirut). South of Lebanon the known localities are numerous: in the Golan Heights and in Upper Galilee this is certainly the most abundant species of *Stenophylax*. In the Golan Heights (18) specimens were caught on Mount Avital, at Ain el Hajal ('Ein el Hadjal), at ca. 1,400 m a.s.l., then at El Rom, at ca. 1,050 m a.s.l., and at Fadjar (Wadi Fajer), ca. 100 m a.s.l., at Senir, and in the Mas'ada forest; moreover, the species is regularly caught on Mount Hermon, as high as 2,000–2,400 m a.s.l., and specimens were taken here several times on the snow, during spring. In Upper Galilee (1) the known localities are Mount Meron, the highest mountain of Israel, up to its top, and H̱uliot (Sedé Neẖemya), only 85 m a.s.l. Finally, Buteicha and Karé Deshe (both 210 m below sea level) are the only known localities in area 7. Most of the numerous specimens collected were taken outside caves between the end of March and the end of May; but I have seen 2 females taken on 25 and 30 November, on Mount Hermon and at Karé Deshe, respectively, 1 male caught on 5 December at H̱uliot, and several specimens taken at Senir (Golan Heights) on 23 December; this is quite a typical picture for a subtroglophile species.

Genus MESOPHYLAX McLachlan, 1882
J. Linn. Soc. (Zool.), 16:157

Type Species: *Limnephila aspersa* Rambur, 1842 (original designation).

Diagnosis: Generally large insects, habitually strongly resembling *Stenophylax*; wings in all respects like in this genus, upper margin of discoidal cell in forewings perhaps slightly less depressed (curved). Spur formula, at least in some species, difficult to ascertain, some of the spurs not being distinct in size or colour from the "normal spines" (see *M. aspersus*); in any event, there is no spur on the male forelegs, or perhaps only a quite minute one. First segment of anterior tarsus longer than 2nd segment. Male genitalia strongly resembling those of *Stenophylax*; at least in two species, including *M. aspersus*, superior appendages small, strongly sclerotized, and characteristically deeply excised (bilobed), and inferior appendages with elongated free part having a moderately complex relief. The female genitalia are more clearly

229

distinct from those of *Stenophylax*, their most interesting characters being: dorsal parts of segments IX and X coalescent or almost coalescent, together forming a large "tubular piece" open distally; ventral part (lobes) of segment IX separated — at least in *M. aspersus* and one other species — by a sharp median keel; supragenital plate markedly concave; and — most reliable character, although subject to intraspecific variation — median lobe of vulvar scale distinctly bifid.

General Remarks: I am firmly convinced that the species presently considered as belonging to *Mesophylax* could be better placed as a species-group in *Stenophylax*. I have retained *Mesophylax* here because there will be no problem in distinguishing the unique Levantine species from all *Stenophylax* using characters of male and female genitalia (deeply excised and strongly sclerotized superior appendages in the male; large "tubular piece" in the dorsal part of female genitalia, and bifid median lobe of vulvar scale). Only 5 species are included in *Mesophylax*, two more or less widely distributed, one — aberrant in several respects — from Madeira, one from Sardinia and one from Ethiopia. Ethologically and ecologically, species of *Mesophylax* strongly resemble those of *Stenophylax*.

Mesophylax aspersus (Rambur, 1842)

Figs. 496–501

Limnephila aspersa Rambur, 1842, *Hist. nat. Névr.*, p. 475.
Stenophylax meridionalis Kolenati, 1848, *Gen. Spec. Trich.*, 1:26, 32, 65–66.
Stenophylax aspersus auct., e.g., McLachlan, 1875, *A Monographic Revision & Synopsis*, pp. 115, 132–133, Pl. 14, figs. 1–6.
Mesophylax aspersus —. McLachlan, 1882, *J. Linn. Soc. (Zool.)*, 16:157 note, 158; McLachlan, 1884, *A Monographic Revision & Synopsis*, 1st Add. Suppl., p. 10; Schmid, 1957, *Trab. Mus. Cienc. nat. Barcelona*, N. ser. zool., 2(2):46–48; Botosaneanu, 1959, *Archs Zool. exp. gén.*, 97 (Notes et Revue, 1):42–44, 48–50; Botosaneanu, 1982, *Bull. zool. Mus. Amsterdam*, 8(22):179–180, figs. 34–39.
Stenophylax aduncus Navas, 1923, *Arx. Inst. Cienc. Barcelona*, 8:158–159, Figs. 5 a–b.

In several ♂♂ from Sinai, wing expanse ranges from 21.5 to 32 mm; in several ♀♀ from Sinai ('Ein Hadjiya) it measures 24.2–25 mm, whereas one ♀ from Mount Meron, Israel, has a wing expanse of 36 mm. The size is known to be very variable in this species. Wings like as in *Stenophylax*; discoidal cell in forewings with upper margin only very slightly depressed, in its distal part only. For various authors, the spur formula is either 0, 3, 4 in the ♂ and 1, 3, 4 in the ♀, or 1, 3, 4 in both sexes, with the mention that the spur of the fore tibia in the ♂ is extremely small; I must say that I was unable to distinguish (from normal spines) either this spur in the ♂, or the subapical spurs in the 2nd and 3rd pairs of legs; apparently thus, in specimens from the Sinai, the spurs are 0, 2, 2 in the ♂, and 1, 2, 2 in the ♀ — a rather odd situation, indeed (they are apparently 1, 2, 4 in the unique ♀ from Mount Meron).

230

Male genitalia: Tergite VIII only with two very small patches covered by very fine spinules. Segment IX laterally very well developed, medio-ventrally very short, dorsally forming a very narrow semicircular ridge. Superior appendages strongly sclerotized, medially concave, in lateral view with characteristic deep excision and therefore bilobed, superior lobe shorter, rounded, simple, inferior lobe slightly longer, of slightly more complex form. Intermediate appendages small, sclerotized, triangular, divergent, apically with small "beak" directed downwards; they arise from vertical sclerotized plates having in their middle a blackened "tooth". In lateral view, inferior appendages very indistinctly delimited from segment IX, free part well developed, almost vertical (slightly oblique), relatively broad, obtuse lower distal angle well protruding, upper median angle with distinct – although short – digitiform projection directed upwards; ventral face of inferior appendage with rather complex relief. Phallotheca (i.e., basal part of phallic complex) attached to internal walls of the gonopods by a pair of very large sclerotized strips; phallus and titillators strongly curved, phallus dorsally trough shaped, titillators more strongly sclerotized than it, apically and subapically with moderately strong spines.

Female genitalia: Segments IX and X forming dorsally, without distinct limit between them, a very large "tubular piece" of characteristic shape, suddenly narrowing in its distal third, and dorso-distally open; this tubular piece is convex dorsally and concave ventrally, where it has a median keel (distinct also in lateral view). Ventral part of segment IX well protruding posteriorly in lateral view, but only slightly protruding on the sides (viewed from above, very little of it is visible on the sides of the "tubular piece"); it is represented by two very slightly concave, setose lobes, separated by a very sharp median keel, proximally split. Supragenital plate strongly sclerotized, roughly trapezoidal but with rounded angles, extremely concave ventrally. Vulvar scale with its three lobes aligned, lateral lobes longer and more slender than median lobe, which is incised apically.

Remarks: *Mesophylax aspersus* is a variable species (see, e.g., Botosaneanu, 1959; Botosaneanu & Gasith, 1971). This is true not only with regard to size, but also to male and female genitalia. The following are a few characters affected by variation: in the male, the size of the spiny patches on tergite VIII, the shape of the gonopods in lateral view, and the number of spines on apex of titillators; in the female there is some variation in the outline of the supragenital plate and especially in the length and apical incision of the median lobe of the vulvar scale. This is quite normal for a species with such a wide distribution. I am, nevertheless, convinced that describing subspecies would be non-sensical, since almost every population differs slightly from all others. The species is closely related to the Western and Central European *M. impunctatus* McLachlan, but quite certainly distinct from it.

Distribution: *M. aspersus*, a successful and expansive species, has an extremely wide distribution along a west-east axis, from the Canary Islands, Madeira and northern Africa to Pakistan and Kashmir. However, its distribution is essentially circum-mediterranean, and it has only a limited distribution in Central Europe; if the northernmost localities are in England (where its occurrence is certainly sporadic and

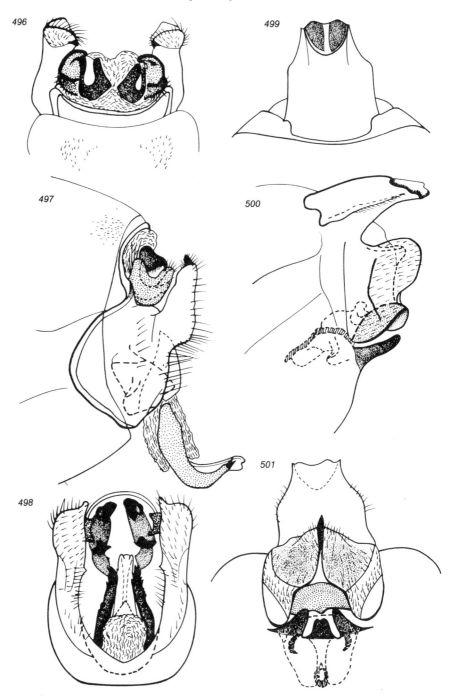

Figs. 496–501: *Mesophylax aspersus* (Rambur, 1842)
496–498. male genitalia (496. dorsal view; 497. lateral view; 498. posterior view);
499–501. female genitalia (499. dorsal view; 500. lateral view; 501. posterior view)

accidental), the southernmost limits of its distribution are apparently reached in the Sinai Peninsula and in Saudi Arabia.

In Israel, a few specimens caught only on Mount Meron (1) and at Nahal Hermon (Banias). Almost all the known Levantine specimens and populations are in the Sinai (22), most of them on Gebel Katharina and Gebel Musa; the localities are: the neighbourhood of the Monastery of St. Katharina, the wadis Taal (Tela), Tlah, and Arba'in (all these wadis at about 1,500–1,700 m a.s.l.). The species was also caught at 'Ein Hadjiya (670 m a.s.l.: 21). There is little doubt that *M. aspersus* will be discovered in all the remaining countries of the Levant.

Ecological Notes: Botosaneanu (1974a: 70–74) published observations on the ecology of this species in the wadis of Gebel Katharina (Djebel Katherine): either permanent or intermittent, more or less rapidly flowing mountain streams in an arid, mountainous environment. *Mesophylax aspersus* — unique representative of the order here — is present in considerable populations, larvae crawling on stones and sand, pupae attached — often in clusters — under the stones. Emergence takes place in spring, very probably during the night, most specimens being caught in Sinai from March to May; *M. aspersus*, like *Stenophylax*, is readily attracted by artificial light. Other specimens were caught in September, after the period of summer quiescence (no Levantine specimen was ever caught during this period in a subterranean hollow, but 1 male was taken in July at Wadi Tlah: a labelling error in this case cannot be excluded).

Family LEPTOCERIDAE Leach, 1815

In: Brewster, *Edinburgh Encyclopedia*, 9, 1 Ent., p. 136 (as Leptocerides)

Type Genus: *Leptocerus* Leach, 1815.

This diagnosis is valid for the subfamily represented in the Levant: Leptocerinae. Moderately small to large insects, more or less slender. No ocelli. Antennae very slender and extremely long (2–3 times longer than forewings, longer in ♂ than in ♀), with long segments, scapus generally bulbous. Maxillary palpi slender, hirsute, with 5 segments, all relatively long, 2nd segment the longest, last segment not annulate (thus not like in Annulipalpia), but incompletely sclerotized and thus flexible (this is sometimes true also for distal part of 4th segment). Legs slender and long; spurs: in forelegs 0, 1, or 2, but always 2 in intermediate and hindlegs. Forewings elongate, slender, with short and dense pubescence. Hindwings shorter than forewings, either narrow and with acute apex, or quite ample, folded; venation moderately simplified, not always easy to interpret, sometimes showing slight sexual dimorphism; in forewing discoidal cell closed, and median cell absent (open); in hindwing discoidal cell open (absent), f3 always absent. Male genitalia complex, with well developed superior and inferior appendages, these last sometimes bisegmented, often with various annex appendages. Apex of female abdomen obtuse, genitalia rather complex, sometimes with structure of segment IX enabling retention of egg-mass.

233

Key to the Genera of Leptoceridae in the Levant

1.   Apical fork 5 absent from hindwing   2
  -    Apical fork 5 present in hindwing   3
2.   Thyridial cell absent from forewing.   **Ylodes** Milne
  -    Thyridial cell present in forewing.   **Adicella** McLachlan
3.   Spurs 2, 2, 2; in forewing long discoidal cell, extending beyond thyridial cell at both ends.
                                                                       Ceraclea Stephens
  -    Spurs 0, 2, 2 or 1, 2, 2; in forewing thyridial cell long, extending basally beyond discoidal cell   4
4.   In forewing, M distinctly furcate at end.   **Setodes** Rambur
  -    In forewing, M apparently not furcate, running straight to wing apex (its lower apical branch — M 3+4 — seeming to arise from the upper branch of Cu).
                                                           Oecetis McLachlan

Notes: There are several imprecise records of Leptoceridae from Israel, most of them insistently pointing to the former presence in this country of the genus *Athripsodes* Billberg.

Tjeder (1946 a: 154) mentions " ... a ♀ Leptoceridae which I hesitate to deal with in the absence of the relative ♂ "; this specimen (labelled "Palestine, E bank of Jordan, near Sea of Galilee, Y. Palmoni") was found in the collections of the B.M. (N.H.), and Dr P.C. Barnard examined it, ascertained that it belonged to *Athripsodes* (the species could not be determined), and prepared the drawings which I include here (Figs. 502–505); I am indebted to Dr Barnard for all this, and for the permission to reproduce these figures.

Tjeder (1946 b: 133) mentions " ... females of two species of Leptoceridae taken by Mr R. Washbourn on the River Jordan, 3.X.1935"; it was impossible to locate these specimens [they are not in the B.M. (N.H.), according to Dr Barnard].

Botosaneanu (1963:95) records "une forme (larva!) avec fourreaux coniques de sable fin" collected in 1940-1944 at H̲ula, and which, according to Flint (1967:77), "may well be a species of *Athripsodes*". In the same publication, Flint mentions, under *Athripsodes* species " ... a single female, apparently of this genus, from Deganya A, 19 April 1965" (I tried to examine this specimen, but it was impossible to locate it).

All this points to the presence, years ago, of maybe two additional species of Leptoceridae, in River Jordan and in the H̲ula marshes, one of them certainly an *Athripsodes*. Presently, there is possibly no leptocerid in the fauna of River Jordan, and the H̲ula marshes no longer exist. The presence of *Athripsodes* in Lebanon is not to be excluded. I did not include this genus in the key; the following non-genital characters will serve to distinguish an *Athripsodes* from a *Ceraclea*: presence of a distinct mid-cranial sulcus (i.e., medio-longitudinal suture on head) and the fact that the 4th segment of the maxillary palpi is entirely sclerotized.

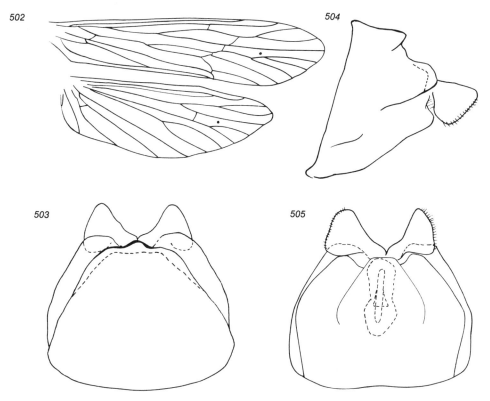

Figs. 502–505: *Athripsodes* sp., female, from Jordan River
502. wings;
503–505. genitalia
(503. dorsal view; 504. lateral view; 505. ventral view).
(Inked from original drawings by Dr P.C. Barnard)

Genus CERACLEA Stephens, 1829
Nomencl. Br. Ins., 1829:28

Type Species: *Phryganea nervosa* Fourcroy, 1785 — a synonym of *Ceraclea nigronervosa* (Retzius, 1783) — by monotypy.

Diagnosis: Larger or smaller, but always slender insects, variously coloured. There is no sulcus (suture) medio-longitudinally dividing the dorsal surface of head. Antennae very long and slender, first segment bulbous. Maxillary palpi with 4th and 5th segments flexible and with a "granulate" shape, 1st segment shortest of all. Spurs: 2, 2, 2. Forewings obtusely rounded at tip, hindwings distinctly broader than forewings in the male, only slightly broader in the female; there is some sexual dimorphism in the venation, too. There is considerable disagreement between various authors concerning interpretation of the venation (e.g., of the apical forks); in forewing,

235

discoidal cell long, sinuate, extending beyond thyridial cell at both ends; no such cells in hindwing; 1st and 4th apical cells (not forks!) petiolate in both wings and sexes. Male genitalia with segment IX neither very long nor very short in its various parts; superior appendages large, free, horizontal, more or less fused at base to form a plate; segment X with main (central) part not divided, forming a horizontal roof, and often accompanied on the sides by pair of slender appendages; inferior appendages variously shaped, but always with large coxopodite — often more or less conspicuously bifurcated, inferior branch arising as projection of baso-distal angle of superior branch — and with very small harpago appended subapically to superior branch of coxopodite; phallic apparatus stout, complex, homologies not well understood, including phallotheca connected by sclerotized stripes to segment IX , membranous endotheca often with sclerotized internal spines (sometimes interpreted as "paramere spines" and apparently replacing the parameres), but in other species with "genuine", external parameres (titillators); in some species a distinct tubular phallus, which cannot be distinguished in other species. Female genitalia having, as most characteristic element, a very strongly developed sternite IX which is flat, forming a large surface which is either undivided, or divided by grooves into 2 or 3 plates.

General Remarks: The species of *Ceraclea* were previously included in genus *Athripsodes* Billberg, 1820 (Morse, 1975; Morse & Wallace, 1976). This is a large genus with some 100 known species distributed in the Palaearctic, Nearctic and — to a lesser extent — Oriental and Ethiopian Regions. The genus is presently divided in 3 subgenera with numerous groups of species. In the Levant only one species is presently known, from Lebanon; it is possible that the genus was earlier represented in Israel, too (see Family Leptoceridae). Ecologically, *Ceraclea* is a diverse genus, but probably most of the species are inhabitants either of stagnant water (lakes especially) or of larger rivers (a good proportion of true potamobionts). The larval cases of *Ceraclea* are often built from mineral particles and cornucopia-shaped: conical, curved, distinctly broader at anterior end than at posterior end, with overhanging antero-dorsal lip.

Ceraclea (Athripsodina) litania Botosaneanu & Dia, 1983
Figs. 506–511

Ceraclea litania Botosaneanu & Dia, in: Dia & Botosaneanu, 1983, *Bull. zool. Mus. Amsterdam*, 9(14):133–134. Type Locality: Litani River near Sir el Gharbii, Lebanon. Type deposited in the Zoological Museum, University of Amsterdam.

Wing expanse of the ♂ holotype: 15.5 mm. Fourth and 5th segments of maxillary palpus with granular structure, flexible. Yellow antennae blackened at the articulations; the darkest part of the body is the dorsal side of the head with its brown-testaceous warts; legs pale, only apex of the first four tarsal segments in all legs distinctly darkened. Spurs 2, 2, 2. Forewings very light brown, almost glabrous. Forewings of ♂ elongate and moderately broad, distally relatively slightly broadened; hindwings broad basally (anal field strongly folding), but curiously narrowed distally. Venation (♂): Fig. 506. In forewings, discoidal cell very long, sinuous, much broader in distal than in proximal half; thyridial cell much shorter, not reaching distal or proximal end of discoidal cell, its root connected by a distinct, oblique cross-vein with inferior edge of discoidal cell (this cross-vein apparently never seen in other species of *Ceraclea*); apical cells (not forks!) 1 and 4 petiolate; there is considerable disagreement between various authors concerning the apical forks; I follow here the opinion of Morse (1975): forks 1, 2 and 5 present; there is no problem with f1 and f5 (f 5 is very angular), but f2 — not accepted by other authors — is delimited upwards by the 3rd branch of SR, downwards by what is interpreted as the 4th branch of SR + anterior branch of M. In hindwings, no discoidal or thyridial cell; apical cells (not forks!) 1 and 4 petiolate; apical forks as in forewings. It is quite probable that the female forewing differs from that of the male especially in featuring one additional branch of M (i.e., in having two branches instead of one between f2 and f5).

Male genitalia: Medio-distally on tergite IX a pair of rounded, contiguous lobes; in lateral view, segment IX considerably longer in its inferior than in its superior part, proximal limit very oblique. Superior appendages forming a plate broadly split in its middle pratically to the base. Segment X with pair of slender latero-ventral supplementary appendages. Inferior appendages with very conspicuous bifurcated coxopodite, and with small harpago appended to the apical and ventral part of its superior branch; superior branch looks like a moderately slender strip obliquely directed upwards and backwards, apical part slightly bent downwards; inferior branch stout, basally very broad but progressively tapering to the obliquely truncate apex; apically this branch has one pair of strong and relatively long sclerotized spurs; in lateral view these spurs slightly upturned, in ventral view strongly divergent, slightly differing in size. Phallic apparatus: the two long internal (endothecal) spurs observed in all species of the *riparia* group, are present, proximal spur simply sinuous, distal spur strongly twisted to form an S (only in dorsal or ventral view).

Female unknown.

Remarks: This species belongs to sg. *Athripsodina* Kimmins, 1963, and to the *riparia* species group which includes some 15 Palaearctic, Nearctic, or Oriental species (Morse, 1975, and later publications). It is certainly the sister-species of *C. riparia* Albarda, 1874, a species known especially from small and large rivers of Central Europe but also from parts of Eastern Europe including zones of the Balkan Peninsula and, maybe, of Anatolia; surprisingly enough, this species was also recorded from southeastern China, and in a letter to the author (11th July, 1988) Prof. J.C. Morse confirmed that the Chinese specimens are *C. riparia*, and not *C. litania*. The two species can be clearly distinguished by several characters of the male genitalia, most distinctive being the much broader base of the inferior branch of the inferior appendages, and the strongly twisted distal endothecal spur.

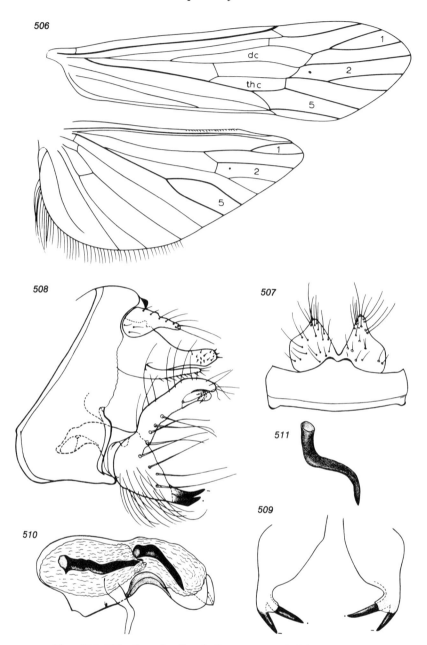

Figs. 506–511: *Ceraclea litania* Botosaneanu & Dia, 1983, male
506. wings;
507–511. genitalia
(507. dorsal view; 508. lateral view;
509. inferior branch of the two coxopodites, ventral view;
510. phallic apparatus, lateral view;
511. proximal internal spur of phallic apparatus, ventral view).
(From Dia & Botosaneanu, 1983, except 506)

Distribution and Ecological Notes: A potential Levantine endemic species, *C. litania* is uniquely known from the Litani River near the village of Sir el Gharbii, Lebanon (ca. 15 km upstream from the river mouth, ca. 200 m a.s.l.). The river here is probably in its Epipotamal zone, the species possibly being an epipotamobiont (but its presence in the Hyporhithral cannot be excluded). The unique known specimen was caught on 14 August 1980. I assume that the species is common in the Litani River.

Genus ADICELLA McLachlan, 1877
A Monographic Revision & Synopsis, pp. 326-327

Type Species: *Setodes reducta* McLachlan (subsequent selection by Kimmins, 1950).
Diagnosis: Habitus variable, some species small and slender, others more robust and larger. Sexes nearly equal in size. Antennae about 3 times the length of wings in the male, rather shorter in the female. Maxillary palpi strong, densely setose. Legs in many species with numerous spines; spurs 1, 2, 2 (spur on anterior tibia short but strong and distinct, spurs on 2nd and 3rd tibiae long, unequal). Anterior wings rather short and broad, with very dense pubescence generally concealing venation; colours variable in different species; hindwings neither much broader nor much narrower than forewings, never folded. Venation not differing in the sexes; in forewings, thyridial cell present and long, much longer than discoidal cell, forks 1 and 5 present (f 1 petiolate), 4th apical cell basally acute, anastomosis often forming a continuous oblique line; in hindwings, no discoidal or thyridial cells, only f1 present, short, anastomosis very short, reduced to one vein connecting SR with M. Male genitalia generally short, not prominent, difficult to describe in general terms (segment IX a sclerotized ring; superior appendages ear-shaped; segment X forming a roof, with or without projections; gonopods either very simple or bifurcated; phallic apparatus generally strongly bent). Apex of female abdomen obtuse, genitalia more uniform than in the male: superior appendages rather large, dorsal part of segment X forming a roof, ventral part represented by two vertical, generally bilobed plates (valves), sternite IX with medio-distal projection between these valves.
General Remarks: Presently slightly more than 30 species of *Adicella* are known; some 15 are western Palaearctic, while 17 are either Oriental or Ethiopian. Although several species groups were recognized, there has, unfortunately, been no attempt at a revision, and doubts were expressed concerning the genus' homogeneity. Some other species may yet be discovered in northern Lebanon, or in adjacent zones of Syria: the males will be readily distinguished from those of *A. syriaca*, but distinguishing the females may represent a problem. All species are elements of the lotic fauna, some of them distinctly limited to springs and small streams, others less restricted in their ecological requirements, and even preferring larger streams. The larval cases are of quite diverse types — pointing to the above mentioned heterogeneity; the cases of *A. syriaca* are conical and characteristically built from elongated vegetal fragments arranged in a very regular spiral; this building pattern is rather unusual for *Adicella*;

the larval cases of the related genus *Ylodes*, represented in the Levant, are also built on this spiral type, and confusion is possible.

Adicella syriaca Ulmer, 1907
Figs. 512–521

Adicella syriaca Ulmer, 1907, *Notes Leyden Mus.*, 29:52–53. Botosaneanu & Novak, 1965, *Acta ent. bohemoslovaca*, 62 (6):477–478. Type Locality: "Syrien, Beyrut". ♂ lectotype (designated by L. Botosaneanu in 1988; an earlier designation of a specimen as lectotype in Botosaneanu and Novak, 1965, is erroneous) kept in the Rijksmuseum van Natuurlijke Historie, Leiden.

Wing expanse, in Levantine specimens: ♂ 12.7–15.8 mm, ♀ 10.8–16.4 mm (Ulmer gives 16–17 mm). Head, thorax and legs pale (yellow, yellowish-brown) with yellowish or white setae: this is the palest European *Adicella* species. Extremely fine and long, pale antennae, distinctly annulated with black. Both wings moderately broad, hindwing only very slightly broader than forewing but shorter and more pointed at apex (apex of forewing parabolic); forewings very setose, setae yellow, but venation distinct in rubbed specimens; anastomosis forming one continuous (oblique) line, very distinct, pale; apical forks 1 and 5 in forewing, where the discoidal and thyridial cells are closed (thyridial cell extremely long, extending much more proximad than discoidal cell); only 1st apical fork present in hindwing, where no discoidal or thyridial cell is present.

Male genitalia: Segment IX forming a complete sclerotized ring; the sternite is the longest part of the segment, whose lateral parts shorten upwards; tergite IX short but medio-distally with small, obtuse projection. Superior appendages relatively large, simple, ear-shaped, with some very long setae. Segment X forming a large roof above phallic apparatus; only the lateral lobes of this segment are present (there is, in other species, a median lobe, too); this segment regularly ogival in lateral view, its distal half medially split by a narrow triangular excision, its two halves (lobes) ending in sharp points; there is, basally on each of these lobes, a characteristic rounded zone of thick chitin covered with minute spinules. Inferior appendages, in lateral view, of very simple shape, elongate, oblique, conical, pointed at apex; in ventral (apical) view they are proximally contiguous, then diverging — with rounded setose projection on median side, finally straight, with capitate spiny apices directed mediad (viewed at a slightly different angle, the apices are more club-shaped than in Fig. 515). Phallic apparatus tightly tied by sclerotized strips to segment IX; in lateral view it is distinctly pistol-shaped; internally there are 3 sclerites almost without doubt belonging to the endotheca: proximally, one pair of long spines apparently curiously bibranched; distally, one much shorter, bifurcate sclerite, shaped like a horseshoe when viewed from certain angles.

240

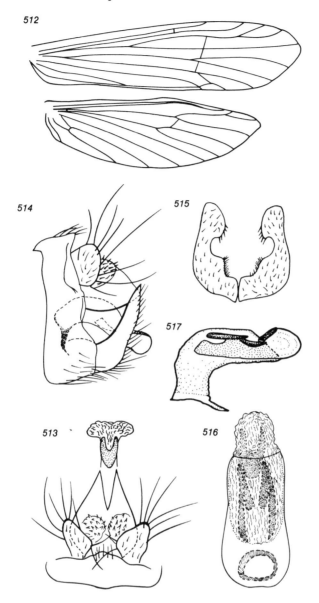

Figs. 512–517: *Adicella syriaca* Ulmer, 1907
512. wings; 513–517. male genitalia
(513. dorsal view, including apex of phallic apparatus; 514. lateral view;
515. gonopods, ventral view;
516. phallic apparatus, dorsal view; 517. the same, lateral view).
(512: from Gasith, 1969;
513–515 & 517: from Botosaneanu & Novák, 1965 —
all but 517 prepared from the type specimen)

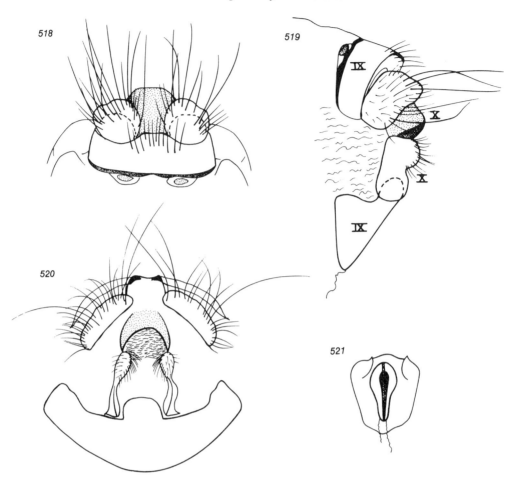

Figs. 518–521: *Adicella syriaca* Ulmer, 1907, female genitalia
518. dorsal view; 519. lateral view; 520. posterior view;
521. vaginal apparatus, ventral view

Female genitalia: Described here for the first time. Tergite IX with very short medio-distal projection, apically truncate (very slightly emarginate, and with pair of extremely faint "tubercles"). Sternite IX longer, with distinct medio-distal quadrangular projection extending between ventral lobes of "valves of segment X". Superior appendages (cerci) large, somewhat irregularly ear-shaped, obliquely placed, with very long setae. Dorsal part of segment X not very long, but longer than superior appendages and therefore quite distinct not only dorsally (where it is truncate distally), but also laterally; it forms a small roof above the genital opening. Ventral part of segment X represented by two important vertical plates ("valves of segment X" or "lateral gonapophyses") which are strongly bilobed, upper lobe with short setae. Vaginal apparatus: a shield as in Fig. 521.

Remarks: *A. syriaca* is relatively isolated, not being closely related to any other known species, at least not to any Palaearctic one. It is only very slightly variable.

Distribution: Known from the Levant, Asia Minor, Greece (including islands like Rhodes and Kithira), Bulgaria, the southern lowlands of Rumania, the Apuseni mountains of Romania, Hungary, Algeria, Morocco; in some of these areas it is a common species, in others its occurrence may be considered as restricted or very restricted. The species was also mentioned from the Caucasus, but this is a slightly doubtful record; it was also reported from eastern Africa, South Africa, and from Ceylon, but in all these cases it is almost beyond doubt that other species were involved (see, for example, Scott, 1986:237). In the Levant, it is a common species in appropriate localities in Lebanon and in Upper Galilee. The Lebanese localities are: "Beyruth" (terra typica); several localities in the catchments of two coastal rivers: Nahr ed Damour (the river itself, and its tributary Nahr el Hamman, 40–45 m a.s.l.) and Nahr el Aouali (the river itself between ca. 50 and ca. 700 m a.s.l.; its tributaries Ouâdi Qachâqich and Ouâdi Râs el Mâ, probably with their sources, at 800 m a.s.l.); the Orontes downstream from its spring Zarka, at Hermel, 650 m a.s.l., as well as several karst springs with their streams, belonging to the Orontes basin (spring Elain, springs and brooks at Labwé; 950–1,000 m a.s.l.); a source and a brook in the karst zone at Baalbek, ca. 1,100–1,150 m a.s.l.; and the upper reach of Yahfoufa, a tributary of the River Litani at 1,200 m. Almost all of the known localities in Israel are in Upper Galilee (1): the rivers Banias, Hazbani, and Dan — being the major water courses feeding River Jordan (ca. 110–380 m a.s.l.); the springs and streams at Tel Dan (Tel el Kadi), at ca. 200 m; Huliot (Sedé Nehemya), at the junction of the three above mentioned rivers (ca. 85 m); the southernmost known locality in the Galilee — possibly also the southernmost point in the species' distribution (!) — is the small stream Nahal (Wadi) 'Ammud, at 100 m below sea level. One known locality in the Golan Heights (18) has to be added: Quosbiya el Jadide, 400 m a.s.l.

Ecological Notes: At many of the localities listed here, fairly large numbers of specimens of *A. syriaca* were caught several times, the populations being certainly sizeable in suitable habitats. The species is a typical creno- and rhithrobiont, inhabiting various types of springs, spring-brooks, larger streams including those belonging to the Hyporhithral, but never the Epipotamal. It is, therefore, a relatively euryoecious species when compared with several other western Palaearctic *Adicella* which are true crenobionts; nevertheless, relatively rapidly flowing, well-oxygenated water is apparently always required. The altitudinal range is considerable: from ca. 100 m below sea level to at least 1,200 m above sea level. On the wing at least from April to the end of November.

Genus YLODES Milne, 1934

Studies in North American Trichoptera, 1:11

Type Species: *Triaenodes grisea* Banks, 1899 (original designation). This should be, according to Manuel & Nimmo (1984), a synonym of *Ylodes reuteri* (McLachlan, 1880).

Diagnosis: Slender, delicate insects, despite their size which is not very small, and their wings which are moderately broad. Antennae very long and slender, 1st segment of male without neoformation (scale with androconial hairs). Spurs 1, 2, 2 (but there is an extremely small spine on the fore tibia, apparently forming a pair with its spur). Forewings moderately broad, elongate, elliptical at end, only very slightly narrower than hindwings, and completely covered by shaggy setae (whitish or greyish, without conspicuous patterns). Venation very characteristic; in forewing: SC and R1 very thick, with anteapical point of contact, discoidal cell large, but thyridial cell absent, because M is represented only by two distal branches directly arising from the almost straight anastomosis, only f1 — petiolate — and f5 present (there is something resembling f2, but this is simply an illusion); in hindwing: SR and M furcate almost at level, no discoidal cell, Cu not furcate (no f5), only f1 present. Male genitalia: segment IX distinctly divided in dorsal and ventral part; segment X well developed as an elongate, oval roof above phallic apparatus; above this roof, a complex of elongate appendages, the central one (unpaired "intermediate appendage") being laterally accompanied by the superior appendages; inferior appendages large, roughly ovoid, apical border deeply excised to form two lobes, internal face concave and with pair of accessory appendages, upper one much longer than lower one; phallic apparatus not connected to inferior appendages or to segment IX, bent, with one long and slender medio-dorsal branch rooted on the phallotheca, and here called "titillator". Female genitalia: Segment X relatively large, laterally accompanied by the relatively small superior appendages (cerci), which are free, roughly triangular; a more ventral position is occupied by a pair of large, setose, vertical "valvae"; between them, the vulvar scale, simple, flat (without a keel), strongly bilobed; vaginal apparatus with long lateral branches, and not accompanied dorsally by a conspicuous "vaginal chamber".

General Remarks: *Ylodes* is a small genus, with 13 known valid species: 9 Palaearctic (some of them with described subspecies of doubtful validity), 2 only in North America, 2 with Holarctic distribution. Most of these species are recorded under the related genus *Triaenodes* McLachlan in various publications (first assigned to *Ylodes* were only two North American species; it was considered by some as a subgenus of *Triaenodes*, but its generic status is presently confirmed).

This genus presents several problems. It is sometimes not easy to distinguish between closely related species. Some species are very variable in their genitalic characters. Occurrence is frequently sporadic, populations often being small and isolated. The habitats, especially of species living in marshy conditions, are destroyed by human activities, and populations of *Ylodes* are sometimes endangered or even considered as extinct.

The (swimming) larvae build conical cases from elongate vegetal fragments very regularly arranged in an elegant spiral, dextral or sinistral.

A key to the two presently known *Ylodes* species from the Levant was considered as being superfluous.

Ylodes internus (McLachlan, 1875)

Figs. 522–525

Triaenodes interna McLachlan, 1875, *Reise Turkestan Fedtcenko Neuroptera*, pp. 35–36, Pl. 3, figs. 3–3b. McLachlan, 1877, *A Monographic Revision & Synopsis*, pp. 322–323, Pl. 34, figs. 1–3; Martynov, 1927b, *Ezheg. zool. Mus (Leningrad)*, 28:462–463, Pl. 19, figs. 7 a–c; Schmid, 1959, *Beitr. Ent.*, 9(3/4): Pl. 10, figs. 8–9.

Triaenodes interna McLachlan ssp. *capitata* Martynov, 1927b, *Ezheg. zool. Mus (Leningrad)*, 28:463, Pl. 20, fig. 2. Tjeder, 1946a, *Ent. Tidskr.*, 67:156–157.

Ylodes internus, Ylodes internus capitatus —. Manuel & Nimmo, 1984, *Proc. 4th int. Symp. Trichoptera* (ed. Morse), Series Entomologica, 30: 220.

Only one (♂) specimen of this species is known from the Levant; it was caught by Y. Palmoni, by artificial light, on 30 May 1937, at Deganya A. ("on the River Jordan, where ... flowing out of the Sea of Galilee"), and it was published by Tjeder (1946) whose excellent drawings of genitalia are reproduced here. I have examined this precious pinned specimen, with genitalia in a Canada balsam preparation, which is kept in the Beth Gordon Museum of Natural History, Deganya, Israel (courtesy: Mr Shmuel Lulav, Director), and I can add a drawing of the phallic apparatus and a few observations.

The mentioned male specimen has a wing expanse of 13.5 mm. Antennae silvery, annulated with brown at intersegmental joints; head and thorax amber-yellow, setae white; forewings completely covered with setae (either white or brownish, forming a mottled pattern) completely concealing the venation.

In particular, two characters of the male genitalia will allow recognition of this species and distinction from the other Levantine species, *Y. reuteri*: (a) the central piece of the dorsal complex of appendages (sometimes called "intermediate appendage"), belonging to segment X, is very long, slender, very distinctly club-shaped at apex; (b) the inferior appendages have a large but relatively slender, parallel sided (margins sinuous) upper lobe, protruding considerably beyond the small, pointed inferior lobe. Other characters are less significant, either because they are subject to much intraspecific variability, or because they are present in other species as well: superior appendages relatively long and slender, but much shorter than "intermediate appendage"; on this last appendage, not far from its base, are rooted two small, club-shaped appendages ("secondary preanal appendages"); main (ventral) part of segment X, forming a roof above phallic apparatus, only slightly shorter than "intermediate appendage", rather slender, quite pointed at apex, with strong

245

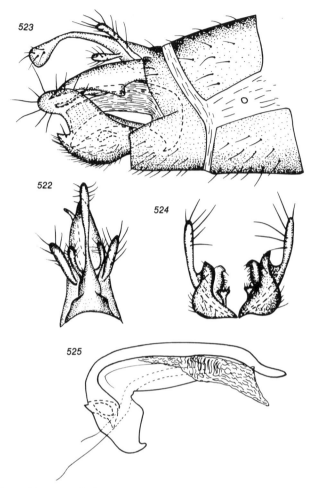

Figs. 522–525: *Ylodes internus* (McLachlan, 1875), male genitalia
522. dorsal view; 523. lateral view;
524. gonopods, ventral view; 525. phallic apparatus, lateral view.
(From Tjeder, 1946a, except 525)

latero-ventral flaps; gonopod with the median complex consisting of two differently shaped annex appendages — one large, one small — observed in all *Ylodes*; phallic apparatus strongly bent, proximal parts sclerotized, distal parts membranous, with single very strong dorsal titillator arising from the phallotheca, distinctly bent anteapically.

The female of this species is known, but I cannot give figures based on Levantine specimens.

Remarks: *Y. internus* is a variable species, this being known from several publications; nevertheless, giving names to the different "forms" does not seem to be a reasonable proceedure. Most closely related to *Y. internus* is *Y. kawraiskii*

246

(Martynov, 1909), but, in the latter species, the apex of the "intermediate appendage" is never club-shaped, and there are also other, minor differences.

Distribution and Ecological Notes: The species has a wide distribution, being known from eastern Siberia (Tomsk), Mongolia, several localities in Central Asia, the plains north of the Caspian Sea, Asia Minor, Iran and Israel. Its occurrence is sometimes apparently sporadic. There are very few reliable data on the species' ecology, but it has been established that it inhabits large rivers (Epipotamal, Metapotamal) as well as lakes. The only known Levantine locality points to its former occurrence in the Jordan River but it is almost certain that no population still subsists in Israel.

Ylodes reuteri (McLachlan, 1880)

Figs. 526–531

Triaenodes reuteri McLachlan, 1880, *A Monographic Revision & Synopsis*. Suppl, 2:65, Pl. 57, figs. 1–2. Martynov, 1909, *Zool. Jb. Abt. f. Syst.*, 27:530, Pl. 26, fig. 30; Tjeder, 1929, *Ent. Tidskr.*, Pl. 3; Botosaneanu, 1963, *Polskie Pismo ent.*, 33(2):98–99; Barnard, 1978, *Entomologist's Gaz.*, 29:244–246.

Triaenodes griseus Banks, 1899, *Trans. Am. ent. Soc.*, 25:214.

Triaenodes reuteri ssp. *turkestanica* Martynov, 1927b, *Ezheg. zool. Mus (Leningrad)*, 28:461, Pl. 19, figs. 6 a–c.

Ylodes grisea —. Milne, 1934, *Studies in North American Trichoptera*, 1:11; Schmid, 1980, *Genera Trich. Canada et États adjacents*, Figs. 684–687.

Triaenodes zeitounensis Mosely, 1939b, *Ann. Mag. Nat. Hist.* (ser. 11), 3:44–46.

Triaenodes simulans —. Kimmins, 1964, *Entomologist*, pp. 40–44 (nec Tjeder, 1929).

Triaenodes reuteri zeitounensis —. Botosaneanu, 1982, *Bull. zool. Mus. Amsterdam*, 8(22):180 (Figs. 41–44: sub *Triaenodes zeitounensis*).

Ylodes reuteri, Ylodes reuteri turkestanicus —. Manuel & Nimmo, 1984, *Proc. 4th int. Symp. Trichoptera* (ed. Morse), Series Entomologica, 30:220.

Three males and one female specimens of this species were caught (April 1942 and June 1943) in the former marshes of Ḥula, in Upper Galilee (1), Israel; three of these specimens, in alcohol and in poor condition, are, for the time being, kept in the Zoological Museum of the University of Amsterdam; these are the only specimens of the species known from the Levant. Botosaneanu (1963) published comments on the species based on these specimens, giving drawings of male genitalia (the same specimens were used for preparing the illustrations of wings and of female genitalia).

In the three Israeli specimens the wing expanse is 15, 16, and 17 mm respectively. Antennae pale, with distinct, narrow, brown annulations at the articulations. The specimens are completely rubbed, the venation (Fig. 526) being quite distinct, especially in a blackened specimen. In particular, two characters of the male genitalia will allow recognition of this species and distinguish it from the other Levantine species, *Y. internus*: (a) the central piece of the dorsal complex of appendages (sometimes called "intermediate appendage"), belonging to segment X, is short (often

247

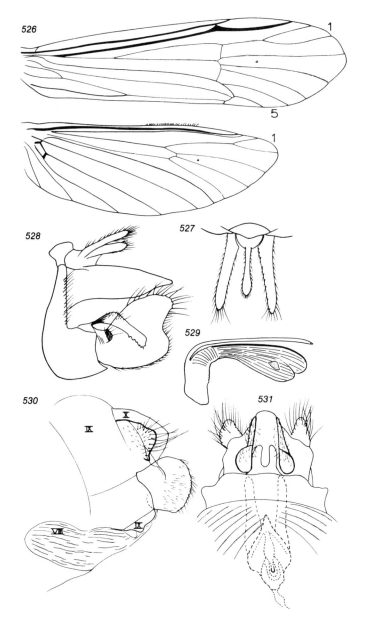

Figs. 526–531: *Ylodes reuteri* (McLachlan, 1880)
526. wings (male);
527–529. male genitalia
(527. dorsal view; 528. lateral view; 529. phallic apparatus, lateral view);
530–531. female genitalia
(530. lateral view; 531. ventral view).
(527–529: from Botosaneanu, 1963)

248

even shorter than superior appendages), in lateral view not club-shaped at apex; (b) the inferior appendages are very broad, both their lobes obtuse, upper lobe not parallel sided, not very strongly protruding beyond large and more or less rounded lower lobe. Other characters are less significant, either because they are subject to much intraspecific variability, or because they are present in other species as well: superior appendages relatively strong, "intermediate appendage" without secondary appendages (but, in various specimens, with a pair of short basal secondary appendages); main (ventral) part of segment X forming a roof above phallic apparatus, much longer than "intermediate appendage", without strong latero-ventral flaps, sometimes emarginated apically (not so in Israeli or Egyptian specimens); gonopod with the median complex of two differently shaped supplementary appendages, their form rather strongly varying; phallic apparatus strongly bent, proximal parts sclerotized, distal parts membranous, with single, slender, dorsal titillator arising from the phallotheca, this titillator apically either straight or slightly curved upwards.

Female genitalia: Figs. 530–531.

Remarks: *Y. reuteri* is a very variable species, almost every part of the male genitalia being affected by this variability; despite a number of published documents, this variability is still very imperfectly known. Some subspecies have even been described; *Y. r. turkestanicus* Martynov is known from Syr Darja, Turkestan; and *Y. zeitounensis* Mosely, described from Zeitoun, Egypt, is certainly not a good species (Botosaneanu, 1963; Manuel & Nimmo, 1984), but it is possibly a slightly distinct geographical race, and the opinion was expressed (Botosaneanu, 1982) that the Israeli specimens may belong to this subspecies, because they share with specimens from Zeitoun the apically non-emarginated segment X and the apically not upturned titillator. Anyway, I presently prefer to place the specimens from Hula only under the specific name. Most closely related to *Y. reuteri* is *Y. simulans* (Tjeder), with a much more restricted European distribution.

Distribution and Ecological Notes: This species is Holarctic in distribution; known from only very few localities in Western and Central Europe, it was found in the British Isles, parts of Scandinavia, the lowlands of Northern and Eastern Europe, the Caucasus and Transcaucasia, Central Asia, Mongolia, Siberia to the extreme east, Egypt, Saudi Arabia, British Columbia, Saskatchewan, Manitoba and Colorado. This distribution is imperfectly known, especially owing to its sporadic occurrence in many of these areas. An inhabitant of standing water and of large rivers, and known also from brackish water, *Y. reuteri* is often found in marshy conditions, and the — unfortunately completely drained — marshes of Lake Hula, in Israel, are an example. It is almost beyond doubt that the species has disappeared from Israel: it was never caught there after 1943, despite repeated attempts, and it is probably absent even from the "Hula Reserve". It was never found in Lebanon; and I am not certain that it still belongs to the Levantine fauna.

Genus OECETIS McLachlan, 1877

A Monographic Revision & Synopsis, 1877:294, 329

Type Species: *Leptocerus ochraceus* Curtis (subsequent selection by Ross, 1944).
Diagnosis: The limits of this genus are still imperfectly known. The variation within it is important, and — in the absence of a much needed revision — I shall limit myself here to a short diagnosis excluding genital characters. Delicate and pale insects, (Fig. 532), with very long and fine antennae, very long and setose maxillary palpi (setae not exceedingly long), spurs 0, 2, 2 or 1, 2, 2. Wings similar in both sexes; forewings elongate and moderately narrow, hindwings either slightly narrower or slightly broader, both wings pointed at apex but not really acute, forewing without distinct anteapical excision of costal border, hindwing with distinct projection of costal border. The most important distinctive characters of the genus are found in the venation of forewing, but some experience is required in order to understand them correctly: forks 1 and 5 present; between these two forks there are 3 veins in the apical part of the wing; of these, the uppermost vein is the posterior branch of SR, the 2nd one (M1+2) directly or almost directly continues the basal part of M, causing this vein to seem unbranched, but, as a matter of fact, M is well branched, its lower branch (M3+4, being the 3rd vein between f1 and f5) appearing to arise from the upper branch of Cu (i.e., of f5) and to form a "supplementary fork"; discoidal and thyridial cells long, sometimes thyridial cell extremely long; anastomosis of quite diverse aspects. In the hindwing, f1 small, f5 larger, basal parts of SR and M normally developed, although rather indistinct, 4th apical cell (not fork!) petiolate, resembling a fork.
General Remarks: A large genus, with probably some 150 known species, well represented on all continents and on some major islands. Many other species will probably be described. I have already emphasized the fact that the limits of the genus are imperfectly known, and that there is much variation within it, having led, for instance, to the description of several genera which were later synonymized. Most of the species are inhabitants either of stagnant water, especially of lakes, or of large and calm running water, but there are exceptions to this, and the known Levantine species is a clear exception. The portable cases of larvae are exceedingly varied with regard to general shape and building materials.

Oecetis terraesanctae (Botosaneanu & Gasith, 1971)

Figs. 532–539

Setodes geissleri Botosaneanu & Gasith, in: Gasith, 1969, Trichoptera of Israel (Thesis, Tel Aviv University), pp. 34–35, etc., Figs. 40–43. Invalid name.
Setodes terraesanctae Botosaneanu & Gasith, 1971, *Israel J. Zool.,* 20:121–129. Type Locality: Tel Dan (Tel el Kadi), Israel. Type kept in the Zoological Museum, Tel Aviv University.

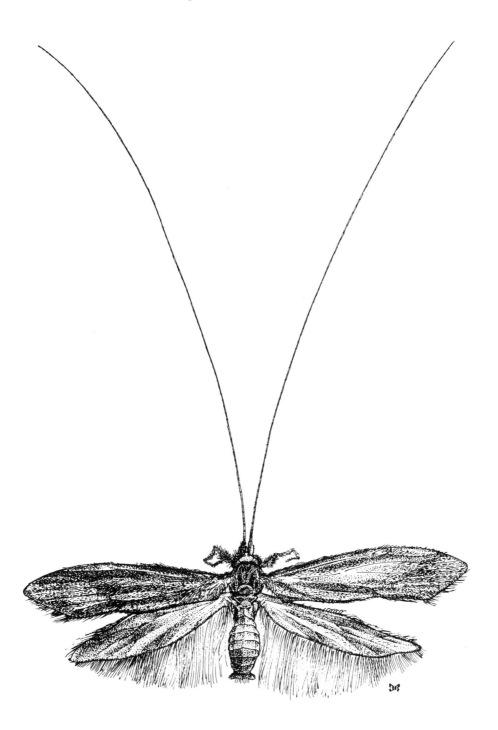

Figs. 532: *Oecetis terraesanctae* (Botosaneanu & Gasith, 1971), female

A very delicate, pale insect (Fig. 532). Wing expanse: ♂ 10.2–12 mm, ♀ 11–11.5 mm. Antennae pale, narrowly annulated with black at articulations, basal segment much shorter than head, without conspicuous tuft of setae. Maxillary palpus very long and setose. Thorax amber-yellow. Spurs: 0, 2, 2; femur, tibia and tarsus of 2nd pair of legs, as well as tarsus of 3rd pair, with numerous spines. Forewing with short, dense, ochreous pubescence, hindwing more greyish, with extremely long postcostal fringe. Forewing elongate, moderately narrow and pointed at apex, clearly broader in its distal than in its proximal half; venation very aberrant, difficult to interpret: SC very thick, strongly sinuous towards its end, and here connected with C by an oblique cross-vein (almost at this level R1 joins SC); SR normal, discoidal and thyridial cells long and of similar length, f1 present and sessile; anastomosis in distinct steps; M arising from Cu very far from root of wing, simply branched, its posterior branch appearing to arise from anterior branch of Cu (thus M seeming unbranched), f5 present, its posterior branch seemingly very short and joining 1st anal vein. Hindwing very narrow, pointed at apex, costal border with distinct projection; part of venation extremely indistinct: SC thickened, joining C just after its projection, and then C thickened until f1; R1 extremely indistinct, like basal parts of SR and M; Cu distinct; 3 anal veins; no discoidal cell; the only forks present are 1 (short) and 5 (longer). On abdomen, in both sexes, a distinct but low medio-dorsal keel.

Male genitalia: Very complex. Segment IX high but not long; tergite distally membranous, here with one or two small lobes, and the small, ear-shaped superior appendages are inserted, too, on this membranous tergal zone; sternite trapezoidal, a large part of it only slightly chitinized. Segment X represented by two pairs of appendages (B and C on Fig. 534) and by an unpaired appendage (A), this complex of appendages having a quite dorsal position; the shape of these appendages is clearly seen in Figs. 533 and 534; appendages B hyaline, appendages C ("intermediate appendages") strongly sclerotized. Inferior appendage very large, most strongly protruding of all parts of genitalia; it is strongly bilobed, proximal (dorsal) lobe large and high, distal (ventral) lobe much longer and more slender, slightly directed upwards and mediad. Phallic apparatus robust, strongly curved, much more swollen in its basal and distal parts than in the middle; sclerotized basal part of phallic apparatus connected by sclerotized bridges to the roots of appendages C (segment X); when in extension, apical part bilobed, lobes distinct in ventral view; no internal chitinous spiniform appendage.

Female genitalia: Characterized especially by the large, flat or slightly concave, oblique sternite IX, forming together with sternite VIII a conspicuous plate. Superior appendages (cerci) rounded, small; segment X very small; ventral lobes (valvae) large, distinctly concave.

Remarks: A very interesting species. In the original publication it was described as a "quite embarrassing Leptocerid". Consideration of all non-genital characters, especially the venation, led to the conclusion that, for the time being, it was preferable to place the species under *Setodes*, pending clarification of the limits of this genus; but similarities with some *Oecetis* were also emphasized. In a letter sent to me (4

August 1983) by Prof. J.C. Morse, he commented, at my request, in the following terms: "Since your description first appeared, I have been impressed with its resemblance to many species of *Oecetis*. On more careful review, I remain convinced of the correctness of that original impression, particularly with reference to the synapomorphous configuration of veins in the M-Cu field in the forewing, wherein the posterior branch of M appears to arise from the anterior branch of Cu, causing M to seem unbranched. Your species belongs particularly to the group of species near *Oecetis* (="*Potamoryza*") *modesta* (Barnard). This is a large group... In this group the long, acute intermediate appendages arise laterally below short superior appendages; most of the species are from the Indian subcontinent or from Africa south of the Sahara".

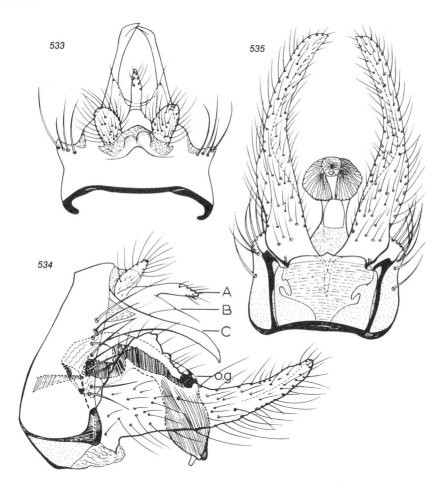

Figs. 533–535: *Oecetis terraesanctae* (Botosaneanu & Gasith, 1971), male genitalia
(533. dorsal view; 534. lateral view; 535. ventral view)
In Fig. 534: A, B, C – segment X appendages; o.g. – genital pore.
(From Botosaneanu & Gasith, 1971)

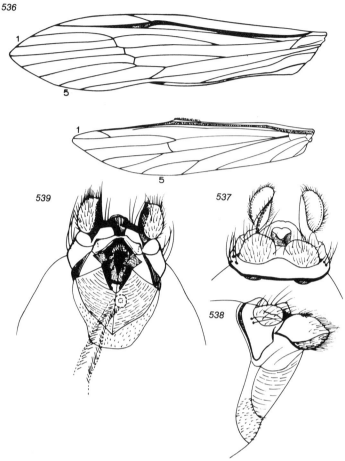

Figs. 536–539: *Oecetis terraesanctae* (Botosaneanu & Gasith, 1971)
536. wings;
537–539. female genitalia
(537. dorsal view; 538. lateral view; 539. ventral view).
(From Botosaneanu & Gasith, 1971)

Distribution and Ecological Notes: This species appears to be a Levantine endemic element. It is presently known from quite a restricted territory, in northern Israel and the Golan Heights; most of the (few) known specimens are from Tel Dan (Tel el Kadi; 1); I have seen (Department of Zoology, Hebrew University of Jerusalem) larvae and pupae collected on 5.X.1981 from Shemurat haHula which is the Nature Reserve of the Hula (1); some imagines were caught on the stream Jalabina (Wadi Jalabina) at ca. 100 m a.s.l., in the Golan Heights (18). Apparently a crenobiont-rithrobiont, but more exact information is needed about its ecology, and the species' presence at Shemurat haHula is rather surprising. Most specimens in the collections were caught in March or April, but the species was also found on the wing as late as early July.

254

Genus SETODES Rambur, 1842

Hist. Nat. Névr., p. 515

Type Species: *Setodes punctella* Rambur, 1842, which is a synonym of *Phryganea viridis* Fourcroy, 1785 (subsequent selection by Milne, 1934).

Diagnosis: Small and delicate leptocerids, always very pale, antennae very long and pale with narrow brown annulations, 1st segment broad basally and tapering distad, not very long, in the male without conspicuous tuft of androconial setae. Spurs: 0, 2, 2. Both wings elongate and narrow, moderately pointed at apex, narrower than in all other Levantine leptocerids; hindwing even narrower than forewing, with posterior margin practically straight, with very long fringes, and with rather well protruding costal angle; forewing with costal margin devoid of notch. Venation in forewings characterized by: SC and R1 independently joining the wing's border; thyridial cell distinctly longer than discoidal cell; f1 (petiolate) and f5 present; important character: M distinctly furcate, its two branches forming a petiolate fork. Venation in hindwings distinct, characterized by: f1 and f5 present, f5 petiolate; important character: SR is not normally developed up to the wing's base, but either evanescent at the anastomosis, i.e., at level of the costal projection (and there connected to M by a cross-vein), or confluent with M at this point; M branches considerably distad from the point where it joins SR or the SR–M cross-vein. Male genitalia: segment IX long ventrally and short dorsally, inferior appendages uniarticulate but complex, bi- or trifid; the large phallic apparatus is a sclerotized tube, strongly curved ventrad, erectile at its tip. Female genitalia strongly sclerotized, massive and complex, segment IX strongly developed, segment X forming a horizontal plate with lateral appendages, broad "ventral valves".

General Remarks: A large genus, almost 200 known species (see especially Schmid, 1987), and present on all continents except South America. Fourteen species are European; Palaearctic Asia, including Japan, has 10 species; there are 8 Nearctic *Setodes*, 4 species in Australia and New Guinea, whereas the Ethiopian Region is inhabited by 14 species; but the Oriental Region is by far the richest of all, with the almost incredible number of more than 130 species.

The systematics of *Setodes* is very complex, and many branches, species-groups, and isolated species, are recognized in the above-mentioned revision, where — fortunately enough! — the unity of the genus is conserved. Two of the Levantine species, *S. viridis huliothicus* and *S. kugleri*, belong to a lineage quite different from that including the 3rd species, *S. alala*.

Most *Setodes* are inhabitants of very different types of running water, only few species being adapted to stagnant water. The larvae build small, solid, conical, slightly curved portable cases, generally from sand.

Key to the Species of Setodes in the Levant

δδ

1. Segment X represented by pair of enormously developed "spines" curved downwards and then forwards; SR and M in hindwing directly connected.   **S. alala** Mosely
–   Segment X represented by short, roof shaped and apically pointed piece; in hindwing, SR connected to M by cross-vein before becoming evanescent   2
2. Superior appendages slender, apically pointed and without brush of stiff, reddish spinules.   **S. viridis huliothicus** Botosaneanu & Gasith
–   Superior appendages strong, apically rounded and with brush of reddish spinules.
     S. kugleri Botosaneanu & Gasith

♀♀

1. Lateral appendages of segment X conical, much shorter than the large segment X; ventrally, segment IX with pair of strong projections.   **S. alala** Mosely
–   Lateral appendages of segment X much longer than the small segment X; ventrally, segment IX without pair of strong projections   2
2. On tergite IX, near root of lateral appendages of segment X, a pair of strong setose warts; the strongly sclerotized "ridges" of segment IX (best seen ventrally: Fig. 545) have a complex conformation.   **S. viridis huliothicus** Botosaneanu & Gasith
–   The warts on tergite IX are minute; strongly sclerotized "ridges" of segment IX (best seen ventrally: Fig. 552) have a simple conformation.   **S. kugleri** Botosaneanu & Gasith

Setodes viridis huliothicus Botosaneanu & Gasith, 1971
Figs. 540, 543–545

Setodes viridis huliothica Botosaneanu & Gasith, 1971, *Israel J. Zool.*, 20:119–120. Type Locality: Ḥuliot (Sedé Neḥemya), Upper Galilee, Israel. Lectotype deposited in the Zoological Museum, Tel Aviv University.

Wing expanse: δ 11–11.5 mm; ♀ 9.5–12 mm. Very delicate insect, all parts of body very pale (the abdomen may be greenish), antennae very narrowly annulate with brown at articulations; spurs 0, 2, 2; tibiae and tarsi of 2nd and 3rd pairs of legs with spines; wings with very long fringes and typical venation; in hindwings, basal part of SR evanescent (SR connected to M by a cross-vein slightly before its evanescence).

Male genitalia: Segment IX (lateral view) very long ventrally, distinctly shortening to the dorsalmost part which is very short; ventro-distal angle truncate. Superior appendages (also called "lateral branches of segment X") very slender, basal part slightly broader, anteapically slightly twisted, only with very few short apical setae. Segment X (lateral view) high basally but tapering to a very slender, elongated and pointed, horizontal apical part. Inferior appendages with two widely separated branches; dorsal branch almost vertical, not markedly broadened, distally more or less clearly bilobed; ventral branch almost horizontal, shorter, moderately broad, with

256

conical projection in the middle of its dorsal face. Phallic apparatus very strongly bent (almost at right angles) in its middle.

Female genitalia (described here for the first time): Lateral appendages of segment X long, relatively slender stripes; near their roots, on tergite IX, a pair of strong, very setose warts. Segment X small, apically not bilobed. The strongly sclerotized, very setose vertical "ridges" on segment IX have a complex conformation, especially when viewed ventrally.

Remarks: In the original description no types were designated. I have subsequently seen the specimens of the type series (some of them deposited in the Tel Aviv Zoological Museum and kindly made available by Dr. A. Gasith), and I have selected a male lectotype (from Hulioth, 28.X.1966) and several male and female lectoparatypes.

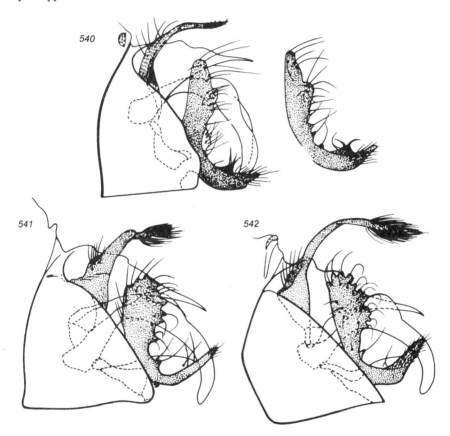

Figs. 540–542: male genitalia, lateral view of three subspecies of *Setodes viridis*
540. *Setodes viridis huliothicus* Botosaneanu & Gasith, 1971
(with the gonopod of another specimen);
541. *S. viridis viridis* (Fourcroy, 1785), from Hungary;
542. *S. viridis iranensis* Botosaneanu & Gasith, 1971, from Iran.
(From Botosaneanu & Gasith, 1971)

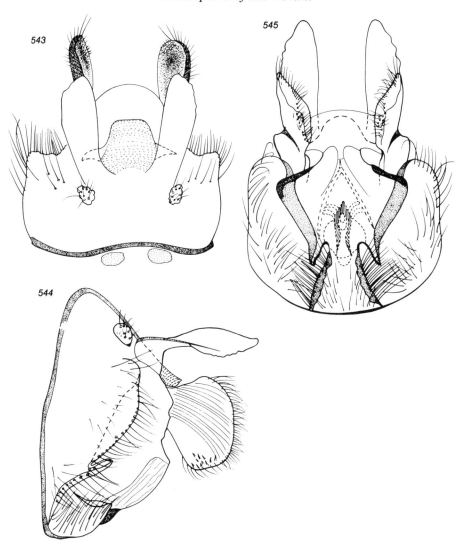

Figs. 543–545: *Setodes viridis huliothicus* Botosaneanu & Gasith, 1971, female genitalia
(543. dorsal view; 544. lateral view; 545. ventral view)

This species has an interesting geographic variability, and presently three subspecies are recognized (Botosaneanu & Gasith, 1971): *S. viridis viridis* (Fourcroy, 1785), *S. viridis iranensis* Botosaneanu & Gasith, 1971, and the above described *S. viridis huliothicus*. A fourth subspecies was described from Bulgaria: *S. viridis bulgaricus* Kumanski, 1976 (Kumanski & Malicky, 1976:120–123); this is considered as a distinct species in Schmid (1987). The most important distinctive characters enabling distinction of these taxa are: lateral shape of segment IX, shape of superior appendages and of the branches of inferior appendages. Illustrations of male genitalia of the nominate subspecies (Fig. 541) and of *S. viridis iranensis* (Fig. 542) are reproduced here. It is quite probable that a similar geographic variability affects the female genitalia.

Setodes viridis belongs to a small, isolated species-group, bearing its name and exclusively represented in the western Palaearctic.

Distribution and Ecological Notes: Most of the few known specimens are from Ḥuliot (Sedé Neḥemya), in Upper Galilee (1), where they were caught either in May or in October; Ḥuliot is at the confluence of the rivers Senir (Ḥazbani), Dan, and Ḥermon (Banias) — three water courses probably belonging to Hyporhithral and forming the Jordan River at ca. 85 m a.s.l. One female specimen from Deganya A (7) was caught in August 1941 and is in the "Beth Gordon" Museum.

Setodes kugleri Botosaneanu & Gasith, 1971

Figs. 546–552

Setodes kugleri Botosaneanu & Gasith, 1971, *Israel J. Zool.*, 20:121–124. Type Locality: Ḥazbani, Upper Galilee, Israel. Type deposited in the Zoological Museum, Tel Aviv University.

Wing expanse: ♂ 11–11.5 mm; ♀ 11 mm. Very delicate insect, all parts of body very pale, antennae with very narrow brown annulations; spurs 0, 2, 2; wings (Fig. 546) with very long fringes and typical venation (in forewing there is a cross-vein between SC and R1); in hindwings, basal part of SR evanescent (SR connected to M by a cross-vein slightly before its evanescence).

Male genitalia: In general terms, as in *S. viridis huliothicus.* segment IX (lateral view) very long ventrally and laterally, but short dorsally, proximal border only slightly sinuous, almost vertical, distal border distinctly convex, ventro-distal angle obtuse; dorsally, between tergite IX and segment X, there is a moderately well developed, less strongly sclerotized zone; sternite IX with straight anterior and rather deeply emarginated posterior border. Superior appendages very characteristic: long, very strong (although not broad), strongly sclerotized; the broader basal part is followed, at an obtuse angle, by a slightly less broad but longer distal part, apex rounded; on median face of the distal half of these appendages, a brush of fine but rigid, reddish spines. Segment X simple, well developed basally, distal part much narrower, ogival in dorsal view, with sharply pointed apex; it is shorter than the superior appendages and very concave ventrally. The inferior appendages may be considered as trifurcated; dorsal branch vertical, club-shaped, borders irregular; ventral branch horizontal and slightly curved, strongly concave dorsally, apex pointed; the median branch is an obtusely triangular and rather small projection located between the two other branches. Phallic apparatus apparently entirely sclerotized, inflated at base, then with horizontal part first narrow and then considerably broadening, finally with long and narrowed distal part directed downwards (and slightly forewards); angle between horizontal and vertical parts almost 90.

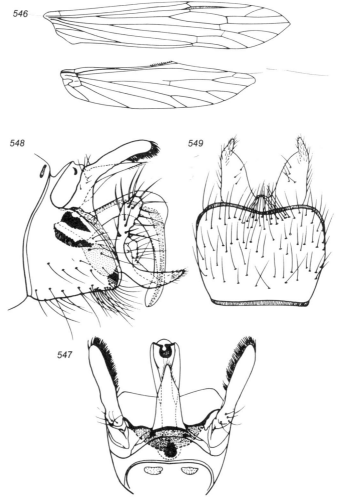

Figs. 546–549: *Setodes kugleri* Botosaneanu & Gasith, 1971, male
546. wings;
547–549. genitalia
(547. dorsal view; 548. lateral view; 549. ventral view).
(From Botosaneanu & Gasith, 1971)

Female: A serious error was made in the original description, especially owing to misleading labelling of some slides: we described and illustrated (Botosaneanu & Gasith, 1971, Fig. 13:1–2) as the female of *S. kugleri* what is clearly the female of *Adicella syriaca*. I have re-examined the female allotype (made available, together with the male holotype, by the Zoological Museum in Tel Aviv; courtesy of Dr A. Gasith). I can presently describe, for the first time, the genuine female of *S. kugleri*, based on one specimen from Litani River, caught together with the male of this species.

Female genitalia: Lateral appendages of segment X relatively short and broad (when compared with *S. viridis huliothicus*); the warts on tergite IX, near their roots, are minute, perhaps with only one seta. Segment X small, apically not bilobed. The strongly sclerotized, not very setose vertical "ridges" on segment IX have a simple conformation.

Remarks: *S. kugleri* certainly belongs to the isolated group of *viridis*; it is closely related to *S. viridis* (Fourcroy).

Distribution and Ecological Notes: Specimens of this species were caught, in very small numbers, on the lower course of Litani River (Lebanon), at Sir el Gharbii, less than 200 m a.s.l.. The type locality is in Israel: Nahal Senir (Hazbani) (Upper Galilee, 1). Specimens were caught in August (Litani) and October (Hazbani). Apparently an inhabitant of the Hyporhithral and Epipotamal.

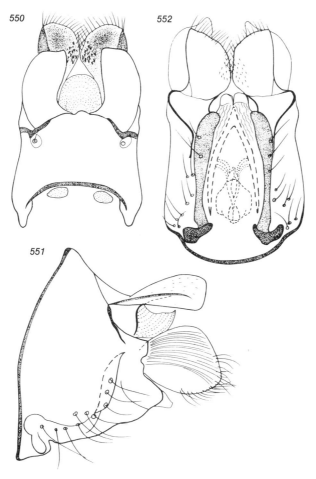

Figs. 550–552: *Setodes kugleri* Botosaneanu & Gasith, 1971, female genitalia
(550. dorsal view; 551. lateral view; 552. ventral view)

261

Setodes alala Mosely, 1948

Figs. 553–560

Setodes alala Mosely, 1948, *British Museum Expedition to South-West Arabia 1937–38*, 1:68–70.
Setodes alalus —. Botosaneanu & Gasith, 1971, *Israel J. Zool.*, 20:119.

In specimens from Sinai, wing expanse of ♂ 10.3–12.5 mm, of ♀ 9.5–11.5 mm, thus males larger than females (for specimens from the Arabian Peninsula, Mosely gives length of anterior wing as 5 mm in the ♂ and 6 mm in the ♀). Very pale insect, antennae silvery with narrow brown annulations. Both wings elongate, narrow; in many cells of forewing, characteristic parallel lines of whitish (silvery) setae on yellowish background and a few spots of contrasting dark setae near apical margin (these white or dark setae readily rubbed off); in hindwing, SR with very short independent trajectory (it apparently originates directly from M, at the level of the costal projection of the wing).

Male genitalia: Posterior border of segment IX, in lateral view, sinuous; medio-dorsally on tergite IX a small but well defined chitinous plate with several very long, stiff setae; in some specimens (e.g., in the holotype: Fig. 554) there is an asymmetric lobe behind this plate (in one specimen I have observed it to be more slender than in this figure; in another specimen there are two symmetric, very small lobes; in most specimens such lobes are absent); sternite IX deeply and broadly excised. Superior appendages very simple, long, baculiform. Segment X represented mainly by a pair of enormously elongate spines, broader at base but suddenly and markedly narrowing, very strongly curved downwards and then anteriad, reaching sternite IX with their tips; these spines are easily seen, being the most characteristic feature of the male genitalia. Inferior appendages also very characteristic, forming a high but not very long complex of two lobes, distal margin of this complex very irregular, with distinct, rounded, but not very deep excision in the middle, separating a superior lobe (broad at base, tapering to a truncate apex with beak-like ventro-distal angle) from an inferior lobe (shorter than the first, quadrangular); from upper basal angle of superior lobe a long and slender appendage projects (this is directed first mediad, then posteriad, its apex clavate; it is not distinct in lateral view, but was represented "en raccourci" in Fig. 557). Phallic apparatus very strongly arched, extending to tips of spines of segment X, slender but with broadened basal and distal parts, distinctly narrowed anteapically, without titillators.

Female genitalia: Segment IX massive, with very irregular distal border, in lateral view; sternite IX with pair of strong and long projections, pointed at apex in lateral view, almost touching at base. Segment X represented by a wide plate, only slightly chitinized, distinctly bilobed apically, lobes rounded; lateral to segment X, a pair of short, conical, widely separated projections. The "ventral valvae" are very wide, medially concave, with rather distinct ventro-distal notch (in this zone, some minute spines on median face). There is also a medio-ventral plate notched at apex ("vulvar scale").

262

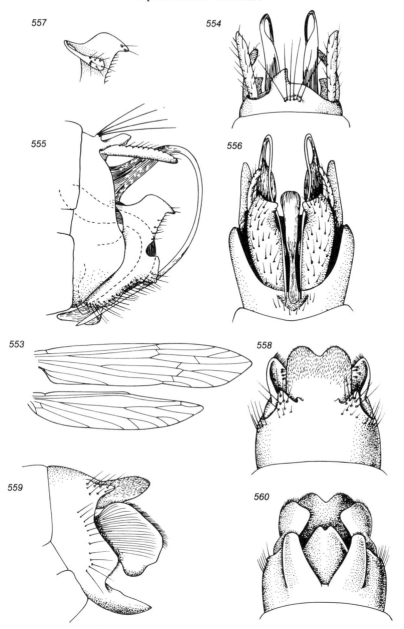

Figs. 553–560: *Setodes alala* Mosely, 1948
553. wings;
554–557. male genitalia
(554. dorsal view; 555. lateral view; 556. ventral view;
557. upper lobe of inferior appendage on inner side);
558–560. female genitalia
(558. dorsal view; 559. lateral view; 560. ventral view).
(From Mosely, 1948; Fig. 553 simplified)

263

Remarks: *S. alala* belongs to a very small group of species. Its sister species is, without doubt, *S. drangianicus* Schmid (Schmid, 1959:777–778, Pl. 9, figs. 9–10), described — as *S. drangianica* — from southern Iran. Botosaneanu & Gasith synonymized *drangianicus* with *alala*, but Schmid (1987) rehabilitated *drangianicus* — certainly a justified decision. Indeed, although very closely related, *S. alala* and the slightly larger *S. drangianicus* can be distinguished by several genital characters (e.g., lateral shape of inferior appendages in the male).

The specific name *alala*, which is not an adjective, was unnecessarily modified to *alalus* by Botosaneanu & Gasith (1971).

Some information on variability of size and of medio-distal projection(s) of tergite IX in the male, was given in the description.

Distribution and Ecological Notes: The description of this species was based on specimens from the former western Aden Protectorate (presently South Yemen: Jebel Harir and Dhala, at ca. 1,600–1,700 m a.s.l.) and from Arabia ("Hejaz: 'Ashara, about 60 miles south-east of Mecca"). In the Levantine Province it was caught on the western shores of the Dead Sea, Israel (13), in the complex of springs and spring-brooks of 'Ein Turabe, ca. 400 m below sea level; moreover, it is known from several localities in the Sinai Peninsula: Wadi Isla, Wadi Watir (22), 'Ein Hadjiya (21); these localities are at ca. 350–700 m a.s.l.). It is a typical "eremial" element, a typical inhabitant of desert watercourses — springs or streams fed by springs, at very different elevations. 'Ein Turabe is possibly the northernmost point in its distribution. Imagines were caught in the Levant either in April, or in August–September; specimens from the Arabian Peninsula were caught either during the first days of May, or in September–November; *S. alala* is possibly a bivoltine species.

Family SERICOSTOMATIDAE Stephens, 1836
Ill. Br. Ent., 6:148, 180 (as Sericostomidae)

Type Genus: *Sericostoma* Latreille, 1825.
Large, heavy insects. No ocelli. Strong sexual dimorphism in maxillary palpi: in ♂ 2–3 segments, but only last segment clearly distinct, adpressed to head, modified or strongly modified to form a "mask"; in ♀ 5 normal segments, 2nd segment longest. Antennae strong, about as long as forewings, with short segments, scapus of ♂ strongly developed and with complex relief. Spurs 2, 2, 4 (or otherwise; but always 2 spurs on foreleg); legs with many dark spines. Wings elongate, apically broadened, apex parabolic, with short, dense setae. Venation similar in both sexes (not in all genera!); in forewing f 1, 2, 3, 5 present, sessile, f1 in basal part completely leaning on upper margin of the closed and mostly angular discoidal cell, thyridial cell very long, no median cell, a very large space between R 1 and SR; hindwing shorter but not narrower, discoidal cell open or closed, f 1, 2, 5. Male genitalia with strong gonopods (basally narrow, distally broadened). Female genitalia with complex relief. One genus represented in the Levant: *Sericostoma* Latreille.

264

Genus SERICOSTOMA Latreille, 1825

Fam. nat. Règne Animal, p. 439 (as *Sericostome*)

Type Species: *Sericostoma latreilli* Curtis, 1834, which is a synonym of *S. personatum* (Spence, 1826) (first included species).

Diagnosis: Large, heavy, dark insects. Head very setose. Antennae shorter than forewings, strong, with short segments, 1st segment in the male very strongly developed, of complex relief, curiously adpressed to dorsal face of head. Sexual dimorphism very accentuated in the palpi; last (2nd? 3rd?) segment of maxillary palpus in the male (Fig. 561) enormously developed into a scent-organ shaped like a "mask" adpressed to the face, convex and with black setae externally, strongly concave on the internal side where a considerable number of pale, woolly setae are developed (maxillary palpus of female consisting of 5 normal segments); labial palpus much longer in male than in female. Spurs: 2, 2, 4. Wings similar in both sexes, covered by thick, short pubescence concealing venation, hindwings with rather long fringes; forewing elongate, distinctly broadened distally, apex parabolic, postcostal border concave, hindwing much shorter, not broader than forewing; in forewing: forks 1, 2, 3 and 5 present, all sessile (f 1, in its basal part, completely leaning on upper margin of discoidal cell), discoidal cell closed, very small and narrow, thyridial cell very long, a conspicuous triangular space between R 1, SR and the anastomosis, root of f 1 connected to R 1 by cross-vein; in hindwing: forks 1, 2, 5 present, discoidal cell open, not connected to R 1 by cross-vein. Male genitalia: segment IX dorsally not delimited from segment X, ventrally very strongly elongate with broad triangular projection extending between ventral branches of gonopods; small, simple superior appendages; segment X medio-apically split, lateral parts ("spines of Xth segment" or "penis sheaths") elongate, strongly sclerotized, ending in some acute projections of various shapes; inferior appendages with very conspicuous dorsal branch (basally narrow, then considerably broadened, apically with deep sinus, median face with two different projections) and very slender and shorter ventral branch (the two ventral branches, distally slightly capitate, forming — ventral view — a characteristic complex); phallic apparatus a long sclerotized tube bent at right angles, with membranous, exsertile part only dorso apically, without titillators or internal spines. Female genitalia: modified tergite VIII (the sternite being smaller than the preceding ones); tergite, sternite, and lateral parts of segment IX represented by well-delimited sclerites of various shapes (Figs. 567–568); "superior appendages": strong, of irregular shape, medially connected to the small but well sclerotized segment X which is apically bilobed, dorsally with parallel sides, ventrally with conspicuous relief and stiff bristles; vaginal apparatus very elongate.

General Remarks: It is generally accepted that the only characters enabling specific distinction in *Sericostoma* are those of segment X of the male (I suspect that other minute differences could be — in some cases, at least — be found in the venation and in other genital parts); however, even the shape of segment X, with its projections, is often subject to important intraspecific variation, and — as a matter of fact — the systematics of this genus is in a permanent "state of flux".

Sericostoma is a purely European (*sensu lato*) genus with moderate effectives: even for the "splitters" it has only slightly more than 20 species, but a few of them are certainly not valid. The presence in the Levant of species other than *S. flavicorne* is improbable. Mention should be made of *Sericostoma mesopotamicum* McLachlan, 1898; I shall not discuss the problem of its validity as a good species (this seems unlikely); the point is that it was described from "Mesopotamia, Malatia", but there is no Malatia in Mesopotamia (instead, there is one in Asia Minor, and a "Malatie" in Egypt).

All *Sericostoma* are running water species, most of them typical rhithrobionts inhabiting streams with a fast current and stony substrate. Larval and pupal cases are large, conical, slightly curved tubes regularly built from sand, and of elastic consistency.

Sericostoma flavicorne Schneider, 1845

Frontispiece, Figs. 561–568

Sericostoma flavicorne Schneider, 1845, *Stettin. ent. Ztg*, 6:155. McLachlan, 1898, *Entomologist's mon. Mag.*, 34:49, Fig.; Moubayed & Botosaneanu, 1985, *Bull. zool. Mus. Amsterdam*, 10(11):63–64.

Wing expanse (Lebanese specimens): ♂ 23.4–26.6 mm, ♀ 26–27 mm. Head, palpi, thorax and abdomen brown or dark brown; antennae + femora, tibiae and tarsi of legs dark yellow (tibiae and tarsi with black spines); in basal part of antenna, a dark line across the segments; membrane of both wings brown, covered with short, dark brown pubescence. Head: Fig. 561. Wings: Fig. 562 (it is possible that slight venation characters distinguish *S. flavicorne* from other species of the genus, but no comparisons are presently possible).

For a general description of male and female genitalia, see genus *Sericostoma*. In male genitalia, the only apparently good specific characters are those of the lateral, sclerotized parts of segment X; despite the slight existing variability affecting the shape and length of the dorsalmost "tooth", the situation is always as shown in Figs. 564–565, and a description would be superfluous.

Remarks: The history of this species is rather complex (Moubayed & Botosaneanu, 1985). It was described by W.G. Schneider based on a specimen from Asia Minor ("Kellemisch"), the description not being illustrated. In the 2nd supplement of his *Monographic Revision and Synopsis*, McLachlan (1880:XLVIII) states that he has seen the type of Schneider. Eighteen years later, McLachlan (1898) writes that he is now convinced that specimens caught at "Beirut, Syria" belong to this species, and he gives the first illustration of the all-important "penis sheaths". All the Lebanese specimens which I have seen agree perfectly with McLachlan's figure, and I am convinced that the Lebanese species is the genuine *flavicorne*. Botosaneanu & Malicky (1978), without having seen material from the Levant or from Asia Minor, decided that the name *flavicorne* — the oldest available name — should be used for a small group of *Sericostoma* "species"; it is presently clear that this initiative, accepted by many, was unjustified: the respective taxa — or some of them — probably belong to only one species, indeed, but another name has to be used for it, *S. flavicorne* being a distinct species probably present exclusively in Asia Minor and the Levant.

Distribution and Ecological Notes: *S. flavicorne* is known from Asia Minor (see also Malicky & Sipahiler, 1984). In the Levant it is known only from the northern and central parts of the Bekaa, catchments of the Orontes and Litani rivers, the localities (ca. 650–1,200 m a.s.l.) being: the Orontes at Hermel downstream from Zarka spring, a tributary of the Orontes at Chlifa, a small stream at Baalbek, the tributary Yahfoufa of the Litani, in its upper and lower course, and karst springs at Ammik, catchment of the Litani. This species is essentially a rhithrobiont, occurring in all the zones of the Rhithral, but it may also inhabit (mainly large) springs.

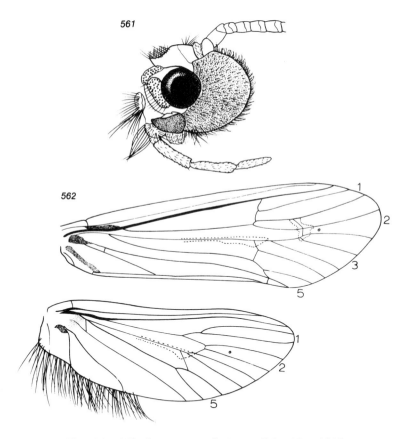

Figs. 561–562: *Sericostoma flavicorne* Schneider, 1845
561. head of male, lateral view, right side;
562. wings

267

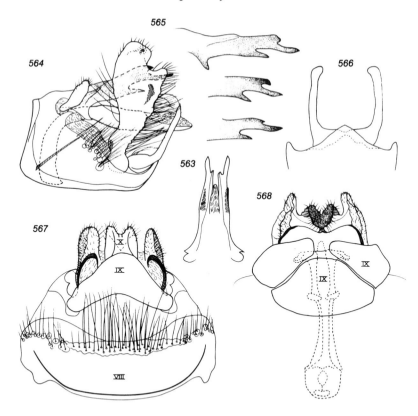

Figs. 563–568: *Sericostoma flavicorne* Schneider, 1845
563–566. male genitalia
(563. segment X, dorsal view; 564. lateral view of genitalia;
565. sclerotized parts of segment X, lateral view, in three other specimens;
566. sternite IX and ventral branches of gonopods, ventral view);
567–568. female genitalia
(567. dorsal view; 568. ventral view)

Family BERAEIDAE Wallengren, 1891
K. svenska VetenskAkad. Handl., 24(10):12, 111

Type Genus: *Beraea* Stephens, 1833.

Small, dark insects, wings densely pubescent. Sometimes with non-genital differences between sexes (e.g., androconial organs in male). No ocelli. Maxillary palpi long, very setose, 5-segmented in both sexes, 1st segment short, 2nd long, 5th not flexible. Antennae rather strong, about as long as forewings, scapus long, thickened. Spurs: 2, 2, 4. Wings in both sexes mostly regularly elliptical, equally broad in both pairs; venation highly simplified and aberrant, slightly different in the sexes, discoidal and median cell absent from both wings. Sternite VII with median "tooth" in both sexes. Male genitalia with very small superior appendages, gonopods unisegmented but with complex annex appendages. Female genitalia: very setose tergite VIII and sternite VIII; an "egg pouch" between sternites VIII and IX; segment X sclerotized, complex.

One genus represented in the Levant: *Ernodes* Wallengren.

Genus ERNODES Wallengren, 1891
(described as subgenus)
K. svenska VetenskAkad. Handl., 24(10):113

Type Species: *Rhyacophila articularis* Pictet, 1834 (by monotypy).

Diagnosis: Small, dark insects, wings densely pubescent. Spurs: 2, 2, 4. In male, no callosity or fold with brush of setae at base of anterior wing, no scales or androconia on either of the wings; no spur on basal segment of antennae, which is normal but long. Maxillary palpus very setose in both sexes, in male 3rd segment abnormally formed, very long and angularly bent (in female normal). Venation simplified and aberrant. There are many non-genital differences between the sexes (see description of *E. saltans*). Abdominal sternite VII in both sexes with median "tooth". Male genitalia: very small superior appendages; segment X forming an "upper penis cover" on both sides of which are strong spines; inferior appendages apparently simple in lateral view, but with complex median armature of lobes; phallic apparatus with strong internal spine of endotheca. Female genitalia: tergite VIII and sternite VIII very setose; between sternite VIII and membranous sternite IX, a large "egg pouch"; segment X complex, well sclerotized, with pair of short but strong lateral appendages.

General Remarks: A small genus, with 9 or 10 described species, all European, only one of them occurring also in Iran (in addition to the Caucasus and the Levant); with the exception of one species with wider distribution in Europe, and one with Central European distribution, the species generally have restricted distributions on some Mediterranean island, in the Balkan Peninsula, the Caucasus, or Asia Minor. They are crenobionts or, at least, decidedly crenophilous. Larval cases are small but hard, conical, slightly curved tubes built from sand; silken membrane closing posterior end with central perforation generally "protected" by some silken but hard cup-like formation.

Certain *Ernodes* species were long considered by some authors as belonging to the related genus *Beraea* Stephens.

Ernodes saltans Martynov, 1913

Figs. 569–577

Ernodes saltans Martynov, 1913, *Rab. Lab. zool. Kab. imp. varsh. Univ. (Arb. zool. Lab. Warschau)*, pp. 51–54, Pl. II, fig. 14; Pl. IV, fig. 10; Pl. VII, figs. 10–11; Pl. VIII, figs. 1–4; Pl. IX, figs. 5, 8.

Wing expanse (specimens from Lebanon): ♂ 8.6–11 mm, ♀ 9.4–11.6 mm. Head and thorax brown, wings with very dense brown pubescence completely concealing venation; head anteriorly protruding (setose tubercle); antennae brown, shorter than forewings, scapus as long as head, with thick dark setae, but of normal form; tarsi spiny (also tibia in intermediate legs), spurs 2, 2, 4 (setose); wings rather broad, of more or less oval form, venation simplified and aberrant; abdominal tergites and sternites distinctly framed. Since many non-genital characters differ in the sexes, an account of them will be given separately for male and female.

Male: Head (Fig. 573) with conspicuous setose wart medially on dorsal surface. Maxillary palpus strongly modified: 2nd segment imperfectly separated from 3rd, with fleshy rounded tubercle externally near apex; 3rd segment very long, elbowed in basal part, with very long and dense black setae in apical part. Mesonotum: scutellum not distinctly delimited, in the middle of mesonotum a pair of very long, pale longitudinal stripes separated on median line by dark space. In forewings, R 1 clearly branched, SR branching once, M branching twice, Cu not branched (thus only one apical fork, the 4th), only one cross-vein (SR-M). In hindwings, no SC, R 1 very thick, SR branching twice, M branching once, Cu not branched (thus only 1st apical fork); there must be a cross-vein SR-M, but it was not represented in Fig. 574. Abdominal sternite VII with median "tooth" much longer than in female, blunt at apex.

Female: Head without setose wart medially on dorsal face. Maxillary palpus with 5 normal, very setose segments. Mesonotum: on scutellum pair of elongate, oval warts, and sometimes also a pair of much smaller, but distinct, warts on anterior part of mesonotum. Forewings: R 1 not branched, SR branched twice, M branched twice, Cu branched, thus f 2(?), 4 and 5 present, a cross-vein SR-M, and another one, very long and sinuous, connecting M and Cu. In hindwings: SC absent, SR and M branching twice, Cu not branched (f 1 and 4 present). Abdominal sternite VII with short, blunt or conical median "tooth" on dark spot.

Male genitalia: Segment IX well developed laterally, in lateral view with important sinus of proximal border in ventral half; dorsally well developed, too (medio-dorsally with rather important proximal excision); sternite IX much shorter; on this sternite is rooted a very characteristic long and strong, slightly asymmetrical dark appendage, tip hooked and turned to the left. Superior appendages very small, setose, placed near roots of intermediate appendages. Segment X quite dorsally placed; its central part is a slightly chitinized short "roof", medio-distally split; on both sides, a characteristic darker intermediate appendage, very broad at base, then spiniform and very strongly

270

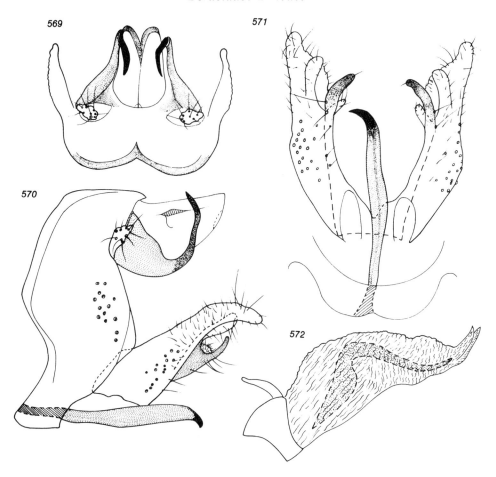

Figs. 569–572: *Ernodes saltans* Martynov, 1913, male genitalia
569. dorsal view; 570. lateral view;
571. sternite IX and gonopods, ventral view;
572. phallic apparatus, lateral view

and obliquely curved upwards (and backwards at tip). Inferior appendages: main, lateral part very elongate, tapering towards the blunt apex, apical part gently bent downwards; on median face of this appendages, a complex of three lobes of different lengths and shapes, not easy to describe (Figs. 570 and 571). Phallic apparatus: basally short, sclerotized phallotheca, followed by ample, membranous endotheca which is trough-shaped (i.e., completely open below) and clearly bilobed apically; from inner wall of endotheca arises a strong, dark spine, broader basally, then very strongly curved and tapering to the apex.

Female genitalia: Sternite VIII proximally telescoped into sternite VII with strong but not very long latero-ventral projections; completely covered with long setae, posterior border not directly continued by IXth sternite: in this manner a large "pouch" is

271

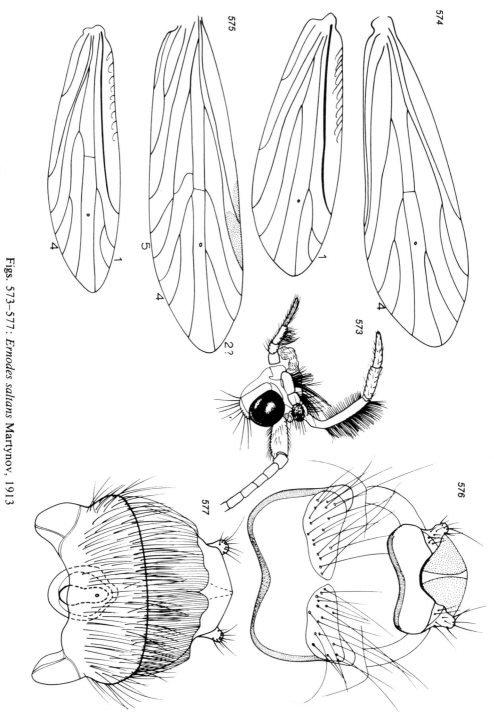

Figs. 573–577: *Ernodes saltans* Martynov, 1913

573. head of male, lateral view, left side; 574. wings of male; 575. wings of female;
576–577. female genitalia (576. dorsal view; 577. ventral view)

formed between the two sternites, certainly for carrying the egg- mass. Tergite VIII distally with two large zones with numerous long setae. In my opinion, segment IX, dorsally as well as ventrally, entirely membranous or slightly chitinized, sternite particularly large, its distal border with several small lobes. Segment X well developed and sclerotized, of complex form particularly well seen in dorsal view (it is not excluded that an XIth segment is also involved here; according to Nielsen, 1980: Fig. 55, the proximal part of this complex, with the lateral "cerci", belongs to tergite IX, the distal part to segment X); the proximal part is a transverse plate, emarginate proximally and more deeply emarginate distally, latero-distally with small but strong setose appendages ("cerci" ?); distal (median) part triangular, apparently divided in two by a pale median stripe.

Distribution and Ecological Notes: *E. saltans* was described from the Caucasus, and rediscovered in northern Iran and in Anatolia.

In the Levant, presently known only from four localities in the catchment of the small Lebanese coastal river Nahr el Aouali (ca. 700–1,000 m a.s.l.): Nahr el Aouali at Jdaîdet ech Chouf; spring Nabaa Mourched of its tributary Ouâdi Qachâqich; spring Nabaa Bâter ech Chouf, also a tributary of Nahr el Aouali, downstream from Niha; and spring Nabaa Aazibi of Nahr Aaray, also a tributary of Nahr el Aouali. This is a crenobiont or, at least, a very distinctly crenophilous species, the same situation being observed by Martynov (1913) in the Caucasus, and by Schmid (1959) in northern Iran. It will certainly be discovered in other suitable habitats in the Lebanese mountains, and also above 1,000 m a.s.l. (in northern Iran it is mentioned also from localities as high as 10,000 ft). Lebanese specimens were caught from the last days of April to the first days of July.

ADDENDA

Family HYDROPTILIDAE
Genus TRICHOLEIOCHITON

Tricholeiochiton fagesii (Guinard, 1879)

Figs. 578–581

Leiochiton fagesii Guinard, 1879, *Mém. Acad. Sci. lett. Montpellier, Sect. Sci.*, 9(2):139–143.

Tricholeiochiton fagesii —. Kloet & Hincks, 1944, *Entomologist*, 77:97.

Dr Heather Bromley and Dr Ch. Dimentman (Jerusalem) sent me, long after the MS of this book was completed, several larvae of this hydroptilid species (with their very distinctive, extremely elongated and slender mid- and hindlegs), correctly determined by them. These specimens were caught in 1941 and 1943 in the former Lake H̱ula and H̱ula swamps (The Hebrew University of Jerusalem Expeditions 1940–1945; Steinitz et al., unpublished data). This is an interesting discovery; the species was never found previously outside Europe. Because it is not impossible that adults will be collected from the Levant, I give here figures of the male.

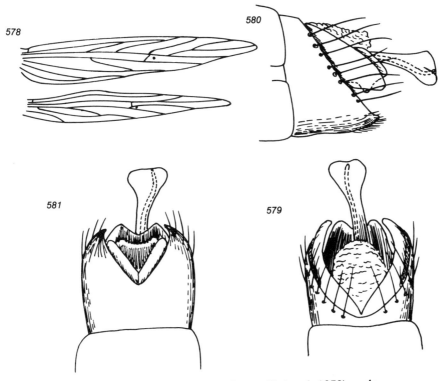

Figs. 578–581: *Tricholeiochiton fagesii* (Guinard, 1879), male
578. wings; 579–581. genitalia
(579. dorsal view; 580. lateral view; 581. ventral view).
(From Mosely, 1939a)

Family LIMNEPHILIDAE
Genus LIMNEPHILUS

Limnephilus turanus hermonianus Botosaneanu ssp. nov.
Figs. 582–588

Type Locality: "Mount Hermon". Type deposited in the Zoological Museum, University of Amsterdam.

Description

Wing expanse: of ♂ 25 mm; of ♀ 29–32 mm. Colour of wings and body pale and uniform: wings very pale, without any markings, antennae and legs uniformly beige with a faint reddish tint, only meso- and metathorax are darker, brown. Apex of forewing not very typical for *Limnephilus*, less sharply obliquely truncate; discoidal cell in forewing almost twice as long as its petiole; no "beard" in hindwing. The leg spurs are very characteristic: in the male, spur formula 2, 2, 2, the spurs on fore legs being very peculiar — black, curved, inserted very near from each other (spurs on the other two pairs of legs normal); in the female there are only two spurs on the middle and hind legs (!), and in the fore legs there is normally one spur, but this is uni- or bilaterally absent in some of our specimens (broken?). There are no "teeth" on some abdominal sternites. Many very short setae on the abdominal sternites.

Male genitalia (Figs. 582–585): Tergite VIII very characteristic; its postero-median part well sclerotized, with almost parallel margins, and bordered by wide fields of membrane, rather strongly protrudes beyond the other parts of the genitalia, having a clearly bilobed apex (the two lobes rounded, covered by minute setae and separated by a sinus as wide as each of the lobes). Superior appendages large and complex, thick, with small median concavity on their posterior face, separated from the rest of the appendage by a "keel" on which is placed, almost in its middle, a strong, black, blunt projection. Intermediate appendages feebly developed (not visible if the genitalia are seen from a perfectly lateral view — which is not the case in Fig. 583), vertical, apical parts blackened. Inferior appendages forming a complex shaped like a lyre, if viewed from behind; their free parts (Fig. 583) are well developed but stout, lower margin angular, distally truncate, each of the distal angles produced into a black point. Phallic apparatus devoid of titillators, phallus more slender in its distal third which is characterized by lateral irregular rows of minute black spinules.

Female genitalia (Figs. 586–588): Tergite IX very strongly depressed (Fig. 586), if viewed laterally much longer at the two ends than in the middle; sternite IX split on the median line into two large plates. Dorsal part of segment X tapering to apex, with very large and deep sinus separating two strong, blackened, triangular (pointed) projections ("superior appendages"?); medio-ventral part of segment X is a large but only slightly protruding plate, distally very slightly emarginate, for the main part pale but with its central part distinctly darkened (Fig. 588). Vulvar scale without peculiarities.

276

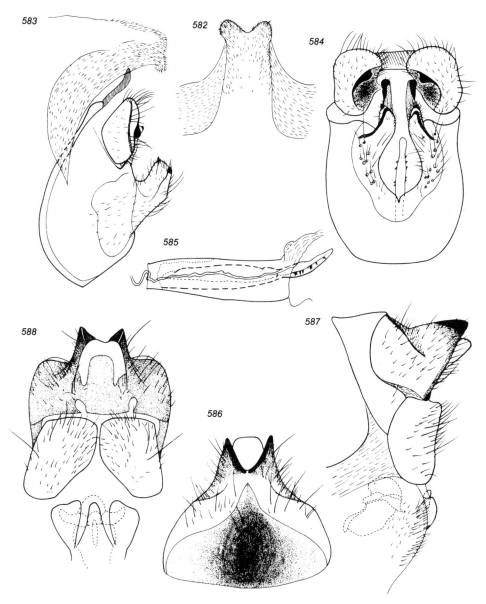

Figs. 582–588: *Limnephilus turanus hermonianus* Botosaneanu ssp. nov.
582–585. male genitalia
(582. tergite VIII; 583. lateral view; 584. posterior view;
585. phallic apparatus, lateral view);
586–588. female genitalia
(586. dorsal view; 587. lateral view; 588. ventral view)

Material and Locality

One ♂ (holotype) and 5 ♀♀ (allotype and paratypes) were caught on 14 December 1987, at a light trap on Mount Hermon (lower station of Mount Hermon trail, 1,600 m a.s.l.) by R. Ortal & G. Müller. Sent to me by R. Ortal (August 1988). Holotype and allotype (pinned, genitalia in vials with glycerin) in the Zoological Museum of the University of Amsterdam; 4 ♀ paratypes, also pinned, in the collection of R. Ortal. *L. turanus turanus* (Mart., 1927) is considered as an inhabitant of mountain lakes; *L. turanus hermonianus* ssp. nov. could be an inhabitant of temporary rain-pools (natural or artificial) in the vicinity of the place where the light trap was operated.

Remarks

Astratodes turanus Martynov, 1927, was described (Martynov, 1927: 488–489, Pl. XXVII, figs. 6–11) from Daghestan (Caucasus) and from Uzbekistan. For this, and for the closely related *iranus*, Martynov erected the new genus *Astratodes*, but Schmid (1955: 141) synonymised *Astratodes* with *Limnephilus*, although considering these two species as forming a species-groups of its own; they are, indeed, distinguished from all other *Limnephilus* by several genital characters, like absence of titillators, but especially by the quite heterodox spur formula, with only two spurs in the middle and hind legs in both sexes, and with the curious pair of black, curved spurs in the fore legs of the male (a character shared with quite a few other species).

From *L. turanus turanus* (Mart.), presently known from the Caucasus, Transcaucasia (Kornouhova, 1986), and Uzbekistan, the new subspecies from Mount Hermon clearly differs in several male genital characters (see also Malicky, 1983: 189): the apical lobes of tergite VIII are less widely distant; and, especially, the inferior appendages are much stouter, with truncate apex with both angles produced into black points (more slender in *L. turanus turanus*, apex upturned, with several black "teeth" on their upper edge). Moreover, Martynov's specimens were larger (wing expanse 32–35 mm).

Limnephilus iranus (Martynov, 1927), known from southern Iran and from the Taurus mountains in Asia Minor (Schmid, 1959: 783), though also closely related to *L. turanus turanus* and *L. turanus hermonianus*, is certainly distinct from them, as shown (male genitalia) by the non-bilobed tergite VIII, the superior appendages devoid of black median projection, the very different inferior appendages in posterior view. It is possible that further research will prove that we have here a complex of three good, biological species, whose variability and distribution are still very imperfectly known.

278

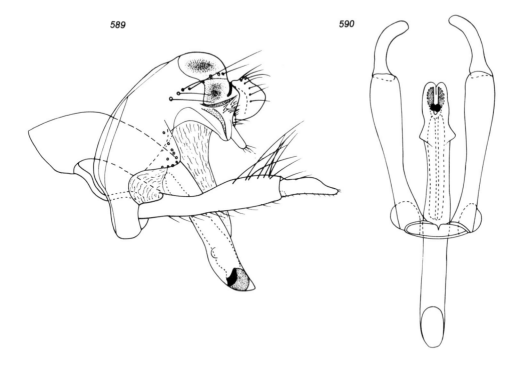

Figs. 589–590: *Hydropsyche instabilis* (Curtis, 1834), male genitalia of a specimen from Naḥal 'Arugot (589. lateral view, 590. ventral view)

REFERENCES

All publications mentioned in the text and older than *A Monographic Revision and Synopsis of the Trichoptera of the European Fauna* by McLachlan with its First Additional Supplement (1874–1884) — more exactly: published before 1884 — are not mentioned in full in these References. More information about them will be found either in McLachlan's Monograph, or in Fischer's *Trichopterorum Catalogus*

Banks, N. (1899) 'Descriptions of new North American neuropteroid insects', *Trans. Am. ent. Soc.*, 25:199–218.

Barnard, P.C. (1978) '*Triaenodes reuteri* McLachlan in Essex, with a note on the identification of the female', *Entomologist's Gaz.*, 29:244–246.

Barnard, P.C. & F. Clark (1986) 'The larval morphology and ecology of a new species of *Ecnomus* from Lake Naivasha, Kenya (Trichoptera: Ecnomidae)', *Aquatic Insects*, 8:175–183.

Botosaneanu, L. (1959) 'Recherches sur les Trichoptères cavernicoles, principalement sur ceux des collections "Biospeologica"', *Archs Zool. exp. gén.*, 97 (Notes et Revue, 1):32–50.

— (1960) 'Sur quelques régularités observées dans le domaine de l'écologie des insectes aquatiques', *Arch. Hydrobiol.*, 56(4):370–377.

— (1963) 'Chrusciki jeziora Houllé (Israel) — Trichoptères du lac Houllé (Israel)', *Polskie Pismo ent.*, 33(2):95–99.

— (1973) 'Au carrefour des Régions Orientale, Ethiopienne et Paléarctique. Essai de reconstitution de quelques lignées "cool adapted" de Trichoptères', *Fragm. ent.*, 9(2):61–80.

— (1974a) 'Notes descriptives, faunistiques, écologiques, sur quelques trichoptères du "trio subtroglophile"', *Trav. Inst. Spéol. Emile Racovitza*, 13:61–75.

— Botosaneanu (1974b) 'Quatre nouvelles espèces palestiniennes de Trichoptères', *Israel J. Ent.*, 9:159–174.

— (1979a) 'Quinze années de recherches sur la zonation des cours d'eau: 1963–1978. Revue commentée de la bibliographie et observations personnelles', *Bijdr. Dierk.*, 49(1):109–134.

— (1979b) 'Deux *Hydropsyche* nouveaux du Jourdain (Trichoptera, Hydropsychidae)', *Bull. zool. Mus. Amsterdam*, 6(21):161–165.

— (1980) 'Quelques Trichoptères nouveaux du pourtour de la Méditerrannée', *Bull. zool. Mus. Amsterdam*, 7(8):73–80.

— (1982) 'Etude de quelques trichoptères ouest-paléarctiques intéressants appartenant au British Museum (Natural History)', *Bull. zool. Mus. Amsterdam*, 8(22):177–188.

— (1984a) 'The Trichoptera of the Levant', in: *Proc. 4th int. Symp. Trichoptera* (ed. J.C. Morse), Series Entomologica, 30, Junk, The Hague, pp. 39–42.

— (1984b) 'Variabilité géographique d'une espèces maghrebino-levantine de *Hydroptila* Dalman', *Ent. Ber., Amst.*, 44:136–139.

References

Botosaneanu, L. & A. Gasith (1971) 'Contributions taxonomiques et écologiques à la connaissance des Trichoptères (Insecta) d'Israel', *Israel J. Zool.*, 20:89–129.

Botosaneanu, L. & J. Giudicelli (1981) 'Observations éthologiques et écologiques sur *Hydroptila hirra* Mosely', in: *Proc. 3rd int. Symp. Trichoptera* (ed. G.P. Moretti), Series Entomologica, 20, Junk, The Hague, pp. 21–29.

Botosaneanu, L. & H. Malicky (1978) 'Trichoptera', in: *Limnofauna Europaea*, 2nd edition (ed. J. Illies), Gustav Fischer, Stuttgart-New York, and Swets & Zeitlinger, Amsterdam, pp. 333–359.

Botosaneanu, L. & M. Marinkovic-Gospodnetic (1966) 'Contribution à la connaissance des *Hydropsyche* du groupe *fulvipes-instabilis*. Etude des genitalia mâles', *Annls Limnol.*, 2(3):503–525.

Botosaneanu, L. & K. Novak (1965) 'Les espèces européennes du genre *Adicella* McL.', *Acta ent. bohemoslovaca*, 62:468–479.

Décamps, H. (1962) 'Note sur quelques espèces de Trichoptères trogloxènes', *Annls Spéléol.*, 17(4):577–583.

Dia, A. (1983) 'Recherches sur l'écologie et la biogéographie des cours d'eau du Liban méridional', Thèse Doctorat d'Etat, Université d'Aix-Marseille III, 302 pp.

Dia, A. & L. Botosaneanu (1980) 'Une *Stactobia* nouvelle du Liban (Trichoptera, Hydroptilidae): ses stades aquatiques et leurs constructions', *Bijdr. Dierk.*, 50(2):369–374.

— — (1982) 'Un cas de gynandromorphisme chez un trichoptère hydroptilide du Liban', *Ent. Ber., Amst.*, 42:140–141.

— — (1983) 'Six espèces nouvelles de Trichoptères du Liban', *Bull. zool. Mus. Amsterdam*, 9(14):125–135.

Fischer, F.C.J. (1960–1973) *Trichopterorum Catalogus*, Vols. I–XV + Index, Nederlandse Entomologische Vereniging, Amsterdam.

Flint, O.S. (1967) 'Trichoptera from Israel', *Ent. News*, 78(3):73–77.

— (1984) 'The Genus *Brachycentrus* in North America, with a proposed phylogeny of the genera of Brachycentridae', *Smithson. Contr. Zool.*, 398:1–58.

Gasith (Geissler), A. (1969) 'Trichoptera of Israel', M.Sc. Thesis, University of Tel Aviv, 82 pp. + Plates + I–II.

Gasith, A. & J. Kugler (1973) 'Bionomics of the Trichoptera of Lake Tiberias (Kinneret)', *Israel J. Ent.*, 8:55–67.

Illies, J. & L. Botosaneanu (1963) 'Problèmes et méthodes de la classification et de la zonation écologique des eaux courantes, considérées surtout du point de vue faunistique', *Mitt. int. Verein. theor. angew Limnol.*, no. 12:1–57.

Jacquemart, S. (1973) 'Description de deux Trichoptères Hydroptilides nouveaux et de l'imago de *Stactobia monnioti* Jacquemart (Ile de Rhodes)', *Bull. Inst. r. Sci. nat. Belg.*, Ent., 49(4):1–16, 2 Pls.

— (1980) 'Un Trichoptère nouveau de l'Aïr: *Hydroptila aïrensis* sp. n.', *Bull. Inst. r. Sci. nat. Belg.*, Ent., 52(13):1–5.

Kelley, R.W. (1984) 'Phylogeny, morphology and classification of the micro-caddisfly genus *Oxyethira* Eaton', *Trans. Am. ent. Soc.*, 110:435–463.

— (1985) 'Revision of the micro-caddisfly genus *Oxyethira* (Trichoptera, Hydroptilidae). Part II — Subgenus *Oxyethira*', *Trans. Am. ent. Soc.*, 111:223–253.

Kimmins, D.E. (1964) '*Triaenodes simulans* in Britain', *Entomologist*, pp. 40–44.

Klapálek, F. (1890, published 1891) 'Nachträge zur Verzeichnis der Trichopteren Böhmens', *Sber. K. böhm. Ges. Wiss. Math.-nat. Kl.*, pp. 176–196.

Kolbe, H.J. (1887) 'Über eine neue, von Herrn H. Tetens bei Berlin aufgefundene Art der Phryganeiden', *Ent. Nachr.*, 13(23): 356–359.

Kornouhova, J.J. (1986) 'Fauna Ruceinikov Kavkaza — The Fauna of Caddis Flies of the Caucasus', *Latvijas Entomologs*, 29: 60–84.

Kumanski, K. (1979) 'The Family Hydroptilidae (Trichoptera) in Bulgaria', *Acta zool. Bulgarica*, 13: 3–20.

— (1983) 'Notes on the group of *sparsa* of Genus *Hydroptila* Dalm., with description of a new species', *Reichenbachia*, 21(2): 15–18.

— (1985) *Trichoptera, Annulipalpia, (Fauna na Balgariia*, 15), Izd. Balg. Akad. Nauk., Sofia, 243 pp.

Kumanski, K. & H. Malicky (1976) 'Beiträge zur Kenntnis der bulgarischen Köcherfliegen', *Polskie Pismo ent.*, 46: 95–126.

— — (1984) 'On the fauna and the zoogeographical significance of Trichoptera from the Strandzha Mts. (Bulgaria)', in: *Proc. 4th int. Symp. Trichoptera* (ed. J.C. Morse), Series Entomologica, 30, Junk, The Hague, pp. 197–201.

Malicky, H. (1970) 'Neue Arten und Fundorte von westpaläarktischen Köcherfliegen (Trichoptera: Psychomyidae, Limnephilidae)', *Ent. Z., Frankf. a.M.*, 80(14): 121–136.

— (1974) 'Die Köcherfliegen (Trichoptera) Griechenlands. Übersicht und Neubeschreibungen', *Ann. Mus. Goulandris*, 2: 105–135.

— (1975, published 1976) 'Beschreibung von 22 neuen Westpaläarktischen Köcherfliegen', *Z. ArbGem. öst. Ent.*, 27(3/4): 89–104.

— (1977) 'Ein Beitrag zur Kenntnis der *Hydropsyche guttata*-Gruppe', *Z. ArbGem. öst. Ent.*, 29(1/2): 1–28.

— (1979) 'Notes on some Caddisflies (Trichoptera) from Europe and Iran', *Aquatic Insects*, 1(1): 3–16.

— (1980a) 'Beschreibungen von neuen mediterranen Köcherfliegen und Bemerkungen zu bekannten', *Z. ArbGem. öst. Ent.*, 32(1/2): 1–17.

— (1980b) 'Ein Beitrag zur Kenntnis der Verwandschaft von *Stenophylax vibex* Curtis, 1834', *Entomofauna* (Linz), 1(8): 95–102.

— (1981) 'Weiteres Neues über Köcherfliegen aus dem Mittelmeergebiet', *Entomofauna* (Linz), 2(27): 335–356.

— (1982, published 1983) 'Köcherfliegen (Trichoptera) von den Kapverdischen Inseln', *Z. ArbGem. öst. Ent.*, 34(3/4): 106–110.

— (1983) *Atlas of European Trichoptera*, Series Entomologica, 24, Junk, The Hague, x + 298 pp.

— (1985) 'Neue Beiträge über mediterrane *Micropterna*-Arten', *Ent. Ges. Basel*, 35(1): 27–35.

— (1986a) 'The Caddisflies of Saudi Arabia and adjacent regions (Insecta, Trichoptera)', *Fauna of Saudi Arabia*, 8: 233–245.

— (1986b) 'Die Köcherfliegen (Trichoptera) des Iran und Afghanistans', *Z. ArbGem. öst. Ent.*, 38(1/2): 1–16.

Malicky, H. & F. Sipahiler (1984) 'A faunistic survey of the Caddisflies (Trichoptera) of Turkey', in: *Proc. 4th int. Symp. Trichoptera* (ed. J.C. Morse), Series Entomologica, 30, Junk, The Hague, pp. 207–212.

References

Manuel, K.L. & A.P. Nimmo (1984) 'The Caddisfly Genus *Ylodes* in North America', in: *Proc. 4th int. Symp. Trichoptera* (ed. J.C. Morse), Series Entomologica, 30, Junk, The Hague, pp. 219–224.

Marlier, G. (1962) 'Genera des Trichoptères de l'Afrique', *Annls Mus. r. Afr. cent., Tervuren, Série in 8°-Sci. Zool.*, 109:1–261.

Marlier, G. & M. Marlier (1982) 'Les Trichoptères de l'ile de la Réunion', *Bull. Inst. r. Sci. nat. Belg., Ent.*, 54(13):1–48, IX Pl., IV cartes.

Marshall, J.E. (1979) 'A review of the genera of the Hydroptilidae', *Bull. Br. Mus. nat. Hist. (Ent.)*, 39(3):135–239.

Martynov, A.V. (1909) 'Die Trichopteren des Kaukasus', *Zool. Jb. Abt. f. Syst.*, 27:509–558, Pl. 24–27.

— (1913) 'K poznaniiu fauny Trichoptera Kavkaza I', *Rab. Lab. zool. Kab. imp. varsh. Univ. (Arb. zool. Lab. Warschau)*, pp. 1–111.

— (1915) 'Zametka o nekotorykh novyh materialah Trichoptera Kavkazsk. Muzeja', *Izv. kavkaz. Muz. (Bull. Mus. Caucase)*, 9:1–17.

— (1916, published 1917) 'Trichoptera of Crimea', *Ezheg. zool. Mus.* (Petrograd), 21:165–199, 369–372.

— (1927a) 'Contribution to the aquatic entomofauna of Turkestan. I. Trichoptera Annulipalpia', *Ezheg. zool. Mus.* (Leningrad), 28:162–193, Pl. VII–XI.

— (1927b) 'Contribution to the aquatic entomofauna of Turkestan. II. Trichoptera Integripalpia, with a note on a new species of *Rhyacophila* Pict.', *Ezheg. zool. Mus.* (Leningrad), 28:457–495, Pl. XIX–XXVII.

— (1934) 'Ruceiniki, Trichoptera Annulipalpia', Tableaux analytiques de la faune de l'U.R.S.S. publiés par l'Institut Zoologique de l'Académie des Sciences, 13:1–343 (Leningrad, Izd. Akad. Nauk SSSR).

— (1938) 'O ruceinikah (Trichoptera) Nahitchevanskoi A.S.S.R. i sopredelnykh stran', *Trudy zool. Inst., Baku (Zakavk. Filiala A.N.)*, 8:65–73.

McLachlan, R. (1874–1884) *A Monographic Revision and Synopsis of the Trichoptera of the European Fauna*; Parts I–IX with Supplements Parts I–II, Appendix and Index: I–IV + 1–523 + I–CIII + 59 Plates; First Additional Supplement: I–IV + 1–76 + 7 Plates (reprint 1968, Classey, Hampton).

— (1895) '*Stenophylax concentricus* Auct. (nec Zett.), renamed *S. permistus*', *Entomologist's mon. Mag.*, 31 (2nd Series Vol. 6):139–140.

— (1898) 'Some new species of Trichoptera belonging to the European fauna, with notes on others', *Entomologist's mon. Mag.*, 34:46–52.

Mey, W. & A. Müller (1979) 'Neue Köcherfliegen aus dem Kaukasus', *Reichenbachia*, 17(21):175–182.

Milne, L.J. (1934) *Studies in North American Trichoptera*, Cambridge, Mass., Vol. 1, pp. 1-16.

Morse, J.C. (1975) 'A phylogeny and revision of the caddisfly genus *Ceraclea*', *Contr. Am. Ent. Inst.*, 11(2):1–97.

Morse, J.C. & J.D. Wallace (1976) '*Athripsodes* Billberg and *Ceraclea* Stephens, distinct genera of long-horned caddis-flies', in: *Proc. 1st int. Symp. Trichoptera* (ed. H. Malicky), Junk, The Hague, pp. 33–40.

Morton, K.J. (1893) 'Notes on Hydroptilidae belonging to the European Fauna, with description of new species', *Trans. ent. Soc. London*, 1893:75–82, Pl. V & VI.

Mosely, M.E. (1930) 'Corsican Trichoptera', *Eos, Madr.*, 6(2):147–184.

— (1933) 'The Trichoptera and Plecoptera of the Auvergne Region of France', *Entomologist*, 66:112–117.

— (1939a) *The British Caddis Flies (Trichoptera), A Collector's Handbook* (illustrated by D.E. Kimmins), Routledge & Sons, London, xiii + 320 pp.

— (1939b) 'Trichoptera collected by J. Omer Cooper, Esq., in Egypt', *Ann. Mag. nat. Hist.*, Ser. 11, 3:43–48.

— (1948) 'Trichoptera', in: *British Museum Expedition to South-West Arabia 1937–8*, Vol. 1, pp. 67–85.

Moubayed, Z. (1986) 'Recherches sur la faunistique, l'écologie et la zoogéographie de trois réseaux hydrographiques du Liban: l'Assi, le Litani et le Beyrouth', Thèse Doctorat d'Etat, Université Paul Sabatier de Toulouse, 496 pp.

Moubayed, Z. & L. Botosaneanu (1985) 'Recherches sur les Trichoptères du Liban et principalement des bassins supérieurs de l'Oronte et du Litani', *Bull. zool. Mus. Amsterdam*, 10(11):61–76.

Navas, L. (1923) 'Excursions entomològiques de l'istiu de 1922', *Arx. Inst. Cienc.* (Barcelona) 8:1–34.

Neboiss, A. (1963) 'The Trichoptera types of species described by J. Curtis', *Beitr. Ent.*, 13(5/6):582–635.

Nielsen, A. (1957) 'A comparative study of the genital segments and their appendages in male Trichoptera', *Biol. Skr.*, 8(5):1–159.

— (1980) 'A comparative study of the genital segments and the genital chamber in female Trichoptera', *Biol. Skr.*, 23(1):1–200.

Nybom, O. (1948) 'The Trichoptera of the Atlantic Islands', *Commentat. biol.*, 8(14):1–19.

O'Connor, J.P. & C. Dowling (1986) 'Notes on some Iranian Trichoptera', *Entomologist's Gaz.*, 37:214–215.

Raciecka, M. (1937) 'Eine neue Trichopterenart aus der Familie Hydroptilidae', *Annls Mus. zool. pol.*, 11(29):477–480, Pl. 56.

Ross, H.H. (1944) 'The Caddis Flies, or Trichoptera, of Illinois', *Bull. Ill. nat. Hist. Survey*, 23(1):1–326.

— (1956) *Evolution and Classification of the Mountain Caddisflies*, Univ. of Illinois Press, Urbana, vi + 213 pp.

Schmid, F. (1947) 'Sur quelques Trichoptères suisses nouveaux ou peu connus', *Mitt. schweiz. ent. Ges.*, 20(5):519–536.

— (1953) 'Contribution à l'étude de la sous-famille des Apataniinae. I', *Tijdschr. Ent.*, 96(1–2):109–167.

— (1954) 'Contribution à l'étude de la sous-famille des Apataniinae. II', *Tijdschr. Ent.*, 97(1–2):1–74.

— (1955) 'Contribution à l'étude des Limnophilidae', Thèse pour l'obtention du grade de docteur ès Sciences, Université de Lausanne, 245 pp.

— (1957) 'Les genres *Stenophylax* Kol., *Micropterna* St. et *Mesophylax* McL.', *Trab. Mus. Cienc. nat. Barcelona*, N. ser. zool., 2(2):1–51.

— (1959) 'Trichoptères d'Iran', *Beitr. Ent.*, 9(1/2):200–219; (3/4):376–412; (5/6):683–698; (7/8):760–799.

— (1960) 'Trichoptères du Pakistan, 3ème partie', *Tijdschr. Ent.*, 103(1/2):83–109.

— (1964) 'Quelques Trichoptères du Moyen Orient', *Opusc. Zool. Münch.*, 73:1–10.

— (1970) 'Le genre *Rhyacophila* et la famille des Rhyacophilidae', *Mém. Soc. ent. Canada*, 66:1–230 + Pl. I–LII.

References

— (1980) 'Genera des Trichoptères du Canada et des Etats adjacents' (*Les Insectes et Arachnides du Canada*, 7; Agriculture Canada), 296 pp.

— (1983) 'Encore quelques *Stactobia* McLachlan', *Naturaliste can.*, 110(3): 239–283.

— (1987) 'Considérations diverses sur quelques genres Leptocerins (Trichoptera, Leptoceridae)', *Bull. Inst. r. Sci. Nat. Belg.* (Ent.), 57 Suppl.: 1–147.

Scott, K.M.F. (1968) 'A new species of *Ecnomus* McLachlan (Trichoptera: Psychomyidae) from South Africa', *J. ent. Soc. sth. Afr.*, 31(2): 411–415.

— (1986) 'A brief conspectus of the Trichoptera (Caddisflies) of the Afrotropical Region', *J. ent Soc. sth. Afr.*, 49(2): 213–238.

Sipahiler, F. (1986) 'Kuzey Anadolu bölgesi *Rhyacophila* (Trichoptera, Rhyacophilidae) türklerinin sistematik yönden incelenmesi' ('A systematic study of the North Anatolian *Rhyacophila* species. Trichoptera, Rhyacophilidae'), *DOGA TU Bio. D.*, 10(3): 524–540.

— (1987) 'Türkiye' deki *Hydropsyche* cinsi *instabilis* grubu (Trichoptera, Hydropsychidae) erkeklerinin sistematik yönden incelenmesi' ('A systematic study on the males of *instabilis* group of *Hydropsyche* genus in Turkey', *DOGA TU Zoaloji D.*, 11(3): 161–178.

Svensson, B. W. & B. Tjeder (1975) '*Oxyethira boreella* n. sp. from Northern Sweden', *Ent. Scand.*, 6: 131–133.

Tjeder, B. (1929) '*Triaenodes simulans* nov. spec.', *Ent. Tidskr.*, 1929: 305–308, Pl. I–III.

— (1946a) 'On a small collection of Trichoptera from Palestine', *Ent. Tidskr.*, 67: 154–157.

— (1946b) 'Trichoptera from the River Jordan, Palestine', *Opusc. Ent.*, 11: 132–136.

— (1951) 'Trichoptera collected in Cyprus by Dr. Hakan Lindberg', *Commentat. biol.*, 13(7): 1–5.

Tobias, W. (1972) 'Zur Kenntnis europäischer Hydropsychidae, I', *Senckenberg. biol.*, 53(1/2): 59–89; (3/4): 245–268; (5/6): 391–401.

Tobias, W. & D. Tobias (1981) 'Trichoptera Germanica. Bestimmungstafeln für die deutschen Köcherfliegen. Teil I: Imagines', *Cour. Forsch.- Inst. Senckenberg*, 49: 1–671.

Ulmer, G. (1907) 'Neue Trichopteren', *Notes Leyden Mus.*, 29: 1–53.

— (1909) 'Trichoptera', in: *Die Süsswasserfauna Deutschlands*, Heft 5&6 (ed. A. Brauer), Gustav Fischer, Jena, iv + 326 pp.

— (1912) 'Trichopteren von Äquatorial Afrika', *Dtsch. Zentralafrika-Expedition*, 4: 81–125 (Berlin).

— (1950) 'Eine neue *Stactobia*-Art und ihre Larve aus Bulgarien, nebst Bemerkungen über die anderen europäischen Arten der Gattung', *Arch. Hydrobiol.*, 44: 294–300.

— (1951) 'Köcherfliegen (Trichopteren) von den Sunda-Inseln (Teil I)', *Arch. Hydrobiol.*, Suppl.- Bd. XIX (Tropische Binnengewässer IX): 1–528 + Pl. I–XXVII.

— (1963) 'Trichopteren (Köcherfliegen) aus Ägypten', *Arch. Hydrobiol.*, 59(2): 257–271.

Wallengren, H.D.J. (1891) 'Skandinaviens Neuroptera. Andra Afdelningen. Neuroptera Trichoptera', *K. svenska VetenskAkad. Handl.*, 24(10): 1–173.

Weaver, J.S. & J.C. Morse (1986) 'Evolution of feeding and case-making behavior in Trichoptera', *J. N. Am. Benthol. Soc*, 5(2): 150–158.

Wells, A. (1979) 'The Australian species of *Orthotrichia* Eaton', *Aust. J. Zool.*, 27: 585–622.

INDEX

Valid names in roman type. Synonyms in italics. Page numbers in bold type indicate principal reference.

286

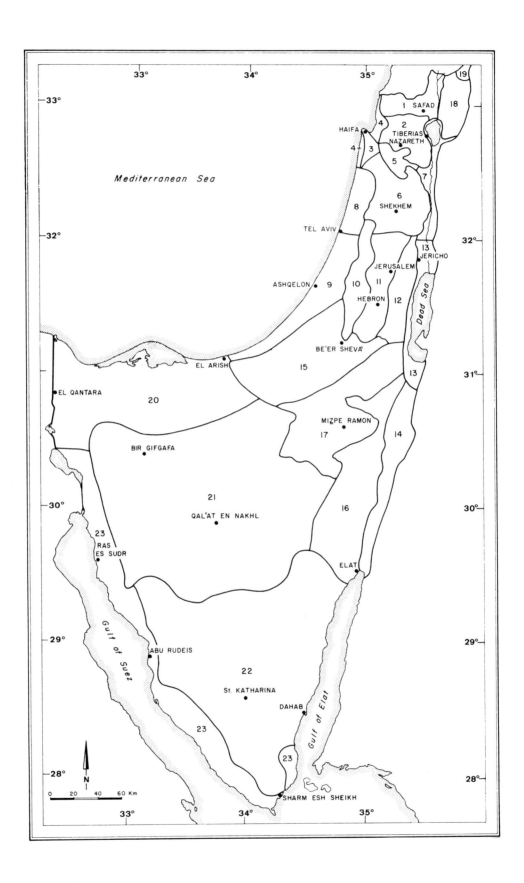

Geographical Areas in Israel and Sinai

KEY

1. Upper Galilee
2. Lower Galilee
3. Carmel Ridge
4. Northern Coastal Plain
5. Valley of Yizre'el
6. Samaria
7. Jordan Valley and Southern Golan
8. Central Coastal Plain
9. Southern Coastal Plain
10. Foothills of Judea
11. Judean Hills
12. Judean Desert
13. Dead Sea Area
14. 'Arava Valley
15. Northern Negev
16. Southern Negev
17. Central Negev
18. Golan Heights
19. Mount Hermon
20. Northern Sinai
21. Central Sinai Foothills
22. Sinai Mountains
23. Southwestern Sinai

כתבי האקדמיה הלאומית הישראלית למדעים

החטיבה למדעי-הטבע

———

החי של ארץ-ישראל

חרקים 6 — שעירי-כנף
של קדמת המזרח התיכון
(TRICHOPTERA OF THE LEVANT)

מאת

לזר בוטושנאנו

ירושלים תשנ״ב